T0123028

MOVING WATER

MOVING WATER

THE EVERGLADES AND BIG SUGAR

AMY GREEN

JOHNS HOPKINS UNIVERSITY PRESS
BALTIMORE

Printed in the United States of America on acid-free paper
2 4 6 8 9 7 5 3 1

Johns Hopkins University Press
2715 North Charles Street
Baltimore, Maryland 21218-4363
www.press.jhu.edu

Library of Congress Cataloging-in-Publication Data

Names: Green, Amy, 1976– author.
Title: Moving water : the Everglades and big sugar / Amy Green.
Description: Baltimore : Johns Hopkins University Press, 2021.
Includes bibliographical references and index.
Identifiers: LCCN 2020019562 | ISBN 9781421440361 (hardcover)
ISBN 9781421440378 (ebook)
Subjects: LCSH: South Florida Ecosystem Restoration Program.
Restoration ecology—Florida—Everglades. | Ecosystem
management—Florida—Everglades. | Sugar trade—
Florida—Everglades. | Sugar growing—Florida—Everglades.
Classification: LCC QH105.F6 G728 2021 | DDC
333.73/1530975939—dc23
LC record available at https://lccn.loc.gov/2020019562

A catalog record for this book is available from the British Library.

Special discounts are available for bulk purchases of this book. For more information, please contact Special Sales at specialsales@jh.edu.

Johns Hopkins University Press uses environmentally friendly book materials, including recycled text paper that is composed of at least 30 percent post-consumer waste, whenever possible.

For Rebecca

WHAT HAPPENS IN THE EVERGLADES is now at a crossroads. Will its sickness, diseased by pollution and thirst, be halted? Can it become a model for sustainable development, with the natural system, farming, and the urban economy adjusting to water management reforms that will make economic development and natural protection more easily achieved?

The history of the Everglades tells us the answer is "no." It will continue to be compromised by those who occupy high government offices, governing but not leading. Environmentalists will rally around it, as they have done for decades, but their voice is not strong enough against the forces that would use the Everglades. Only a swelling cry of citizen support will change the dynamics of the Everglades, from systematic destruction to restoration. . . .

The Everglades uses no power, creates no pollutants, and if brought back to life will yield sustainable benefits forever.

GEORGE BARLEY
Chairman, Save Our Everglades
1994

CONTENTS

CONTENTS

ABOUT THIS BOOK

This book is a work of journalism, meant as a record of pivotal events in Florida history that stand to shape the state for generations to come. Although the book is about the Barleys, their views do not necessarily reflect mine. My job as a journalist is to present the facts as clearly and eloquently as I can and let you, the reader, decide your truth.

I spent some ten years researching this book, which is based primarily on news articles, the Barleys' personal records, and scientific papers. It is also based on countless interviews I conducted and email exchanges with academics, advocates, politicians, and other experts, as well as farmers and residents of the Everglades Agricultural Area, many of whom—despite their helpfulness—asked that their names not be included. I have respected their wishes even as I have tried to fairly include their perspectives.

Where possible I have tried to corroborate events using multiple sources. Dialogue is taken from news articles and interviews as it was recounted to me. These words belong to the individuals who uttered them, and I have not altered them.

MOVING WATER

INTRODUCTION

At the heart of everything is the water.

The water is brown but not muddy. Perhaps this surprises you. Perhaps you imagined mud. Thick, dense mud. Stagnant, suck-the-shoe-off-your-foot mud. Instead the water is clear. Peer into it, and you can see fish dart and turtles paddle. Often the water is described as tea-colored because, like tea, it is colored by leaves, most notably from cypress and pine trees. The water is cool, refreshing even on a hot day. It flows, an important fact for the Everglades. Sometimes the water is smooth as a mirror, reflecting the image of the verdant beauty surrounding it. Tall, leafy cypress trees. Shorter, stouter pond apple trees. When this occurs, you do not peer into the water but stand back and admire the Impressionist painting the water has created.

I once heard the famed photographer Clyde Butcher, known for his stark, black-and-white images of Florida's wilderness, say he drinks the water right from the swamp. Imagine Santa Claus in the swamp and you can imagine Butcher, water up to his wide girth, shirt sleeves cuffed at his elbows, brimmed hat shading his white beard.

"We've drank out of the swamp, and I'm still here," he said. I followed Butcher into the swamp, water reaching my hips. "This is so cool," I said. But I did not drink the water.

Moving water is tricky. By this I mean the feat of relocating water from one place to another. Shortening and straightening rivers. Draining lakes into the sea. The engineering is easy. We've done that for two centuries. But moving water is more complex than the fundamental task of disposing of it to make space for new homes and ignite a housing and agricultural boom, which is what early Floridians envisioned for South Florida and what they got. Moving water sends a cascade of consequences throughout an ecosystem as unique and diverse as the Everglades, some anticipated and some not.

When the Old Ingraham Highway opened in the early 1920s, it was the first to give Ford Model Ts access to the fishing village of Flamingo on Florida Bay near the peninsula's southernmost tip. Hunters and moonshiners used the highway during the dry season as a route into the Everglades. During the rainy season the highway was often impassable. Parts of the highway were constructed on limestone dug from the sawgrass prairie by a dredge floating on barges. The Homestead Canal left behind by the dredge flowed parallel to the highway. Today remnants of the highway are part of the Anhinga Trail in Everglades National Park. The trail is among the park's most popular, offering abundant views of wildlife including alligators, turtles, anhingas, herons, egrets, and many other birds, especially during winter.

After the highway opened Floridians noticed a change in the vegetation on the highway's south side, where it crossed Taylor Slough. Carolina willow and pond apple flourished, while on the highway's north side the sawgrass prairie remained. Floridians would discover that the highway served as a dam for the river of grass, altering the flow of water subtly but enough to alter the vegetation. Scientists still are not completely sure how this happens, says Tom Van Lent, senior scientist at the Everglades Foundation. They believe somc of it has to do with nutrients. The Everglades are a low-nutrient ecosystem. The water flows slowly over a large land area, spreading nutrients thinly; and it may be that the Old Ingraham Highway concentrated nutrients in certain places, transforming the vegetation.

In an ecosystem as unique and diverse as the Everglades it is possible we have not yet experienced all of the consequences of moving water, and we can only wonder about the future. Sometimes our perception of consequences changes. Our values change, and we come to hold dear things we once hardly valued at all. A half century after the Everglades were drained, shrinking the river of grass by half, the region's population has soared far beyond what early Floridians imagined. Demographers anticipated 2 million people in South Florida by the turn of the century. Today 8 million people live there, and forecasts are for the massive influx to continue. The population has grown so fast that South Floridians are realizing they could run out of water very soon.

"It may also be," Van Lent says, "some things happened that we didn't even notice happen, and they're irretrievably gone."

Everglades restoration means, again, moving water, and this is more complex than the fundamental task of acquiring land and removing dams. Restoration will lead to new consequences, some anticipated and some not. It is possible some of these consequences will require new acts of salvation, which in turn will lead to new consequences.

I MET MARY BARLEY IN 2008, a tumultuous time as the world economy was collapsing and the United States was about to elect its first Black president.

I was reporting on the latest development in one of the world's most substantial efforts at ecological restoration. Florida's governor had announced a $1.75 billion plan to buy out U.S. Sugar Corp., the nation's largest producer of sugarcane, and put the land toward Everglades restoration. For generations sugar growers have worked the black earth south of Lake Okeechobee where the river of grass has been refashioned into millions of rows of sugarcane. The governor's plan, bold as it was at the time, offered a huge advancement in decades of acrimony over the Everglades, which had frustrated multiple governors, influenced national elections, and given rise to a massive restoration. The governor's idea: rather than move water, move the thing that is causing Floridians to move the water.

I undertook an assignment for *Newsweek* on the Everglades Foundation, the powerful, little-known organization behind the governor's plan. I phoned Mary Barley, the foundation's chairwoman. Her husband, George Barley, had established the organization before perishing in 1995 in a plane crash while on his way to meet with the US Army Corps of Engineers about the Everglades. Mary described the plan as "the most stunning news since we started working on Everglades restoration." I asked for some background information on the foundation, and in the space of two typed, single-spaced pages of notes Mary told her own story, which is also the story of the foundation.

Weeks later I was on my way to Mary's Islamorada home, where I would spend two days on assignment for the *Christian Science Monitor* interviewing Mary in Islamorada and both her and Tom Van Lent in Everglades National Park. The drive from Orlando where I live is five hours, down the Florida Turnpike and through the Florida Keys on US 1. I listened in the car to news of the unfolding financial crisis. Lehman Brothers had filed for bankruptcy the day before. The presidential election was weeks away.

"The fundamentals of our economy are strong," Republican John McCain declared from the campaign trail in Florida, and Democrat Barack Obama rushed to criticize his presidential rival for seeming a little out of touch with reality.

"He doesn't get what's happening between the mountain in Sedona where he lives and the corridors of Washington where he works," Obama charged. "Why else would he say, today of all days, just a few hours ago, that the fundamentals of the economy are still strong? Senator, what economy are you talking about?"

McCain responded by softening his stance on the economy's strength, explaining that while it was true the fundamentals were "threatened" and "at risk," his previous statement had been meant as praise for the resilience of American workers.

"My opponents may disagree, but those fundamentals—the American worker and their innovation, their entrepreneurship, the small business—those are the fundamentals of America, and I think they're strong," McCain said from Orlando.

I arrived in Islamorada in the early afternoon. I turned right off US 1, heading in the direction of Florida Bay and Barley Basin, named for George Barley years before during a festive ceremony. I drove beneath a pink canopy of blooming bougainvillea and parked my car in a gravel drive. I slammed the car door shut and left behind the financial crisis and presidential campaign. I walked past a shaded courtyard decorated with Buddhist ornaments and enclosed by a bamboo fence, where 5-inch-by-7-inch framed paintings hung. The paintings were the kind tourists buy for a few bucks, kitschy depictions of those things tourists love most about the Florida Keys. Roseate spoonbills, for instance, birds pink as flamingos. I rang the bell beside two large copper doors. Off to my right I could see Barley Basin, the same shade as the sky. Mary came to the door with a cacophony of Havanese, Shih Tzu, and Maltese-mix dogs. I was shown to Mary's office, where we would have our first interview of my trip, in a small separate building of her 2,000-square-foot home, above a guest room. The office was small with a grand view of a pool, tiki bar, hammock, small boat and dock, and Barley Basin. Beyond the basin is Florida Bay and Everglades National Park, which Mary can see on bright days.

Mary wore a short-sleeved linen-colored shirt with cuffed sleeves, khaki slacks, and a large red beaded necklace. Her caramel-colored hair was cropped at her shoulders. Her manner was warm but guarded, direct, no-nonsense and intense. You didn't immediately notice her 5-foot-1 stature. I could tell she was not your average environmentalist. She offered me a plastic, disposable bottle of water.

"Would I go and sit on top of a tree?" she said. "Probably not, because I know I could be more effective doing other things."

Mary Barley explained that her and her husband's Everglades advocacy was as much about addressing what they believed was government corruption as about saving the environment. She recalled a childhood of poverty in Wisconsin, where she grew up one of five siblings raised by a single mother on government assistance. She described her and George's lobbying in Washington and Tallahassee and later her campaign for the office of commissioner of the Florida Department of Agriculture and Consumer Services.

"I think in the beginning we were pretty naive," she said. Government leaders "are supposed to be servants of ours, and we are the ones who pay them, and I came from a very high-minded family, and we thought government did the right thing, and that was how we were raised. . . . They made roads, and they built schools, and they tried to help people like my mother who, we could not have survived without assistance. . . . It's hard to really realize that that's not how it is. The government is run by special interests and big contributions, and the bigger contributions you give the more likely you are to get what you need or what you want."

I asked Mary about her nastiest political experience. She told me about an ad that had aired during her campaign for agriculture commissioner. The ad described her as "barely a Democrat."

"I think it was the nastiest because, what did that have to do with what I was doing?" she asked. "I think that sticks in my mind because it was effective. It kind of shows you how uninformed the public is about the issues instead of the personality. I see that a lot in what is happening right now. The issues are so different than the personality, and people want to concentrate on that part of you and not on what's good for them. It still amazes me to this day, now that I know more about politics, how little the public knows about what they should vote for themselves. They can be swayed really by the most silly arguments. What difference does it make? If you want good public policy you should [elect] the people who are willing to do the right thing, whether they're Republican or Democrat. It should be about who's going to do what's right for the public and have the willpower to go beyond the special interests. . . .

"Within that context is, you're out there trying to get the government to do the will of the people, and that's hard for people to understand. They always think if you're doing something you're doing it for yourself first. . . . That's not my gig. Of all the places I thought I'd end up in my life it was not politics. . . .

"Sometimes I think in profound ways about why it is so hard to get people to want to help and why don't they understand. What is it that we're missing to get them to understand that if we can save [the Everglades] we can probably save any place. It's just our political will, and we don't really have it because we have so many messages that the public doesn't even know what's good for them anymore. They're just like in this, tossed around in this washing machine going, OK, what's important today?"

I asked Mary whether she enjoyed politics.

"It's first a big responsibility, and then I've been blessed and I need to give back, and part of that, of giving back, is giving back to what made you feel blessed, and to me that is the state of Florida," she said.

"I feel very comfortable in trying to do that."

1

GEORGE BARLEY'S BIRTHDAY

The Florida Keys trail the peninsula's southernmost tip like paint dripped from God's heavy brush, extending the east coast in a broken line protruding beyond the point where the east coast and west coast are supposed to meet. Situated between the Keys and the peninsula's southernmost tip, where Everglades National Park is located, is Florida Bay.

Florida Bay is the shimmering teal smeared with purple you see off to your right as you drive south away from Miami on US Route 1, the Overseas Highway, which threads the Keys and tethers them to the peninsula. Beneath almost unceasing sunshine, manatees, sea turtles, stingrays, trout, barracudas, sharks, and dolphins wearing never-ending smiles fill the bay like an aquarium. Game fish like tarpon taunt fishermen like George Barley. Tiny islands—dozens more paint drips—dot the bay, sprouting mangroves, trees that seemingly hover above the water, their roots visible briefly before shooting beneath the surface. These tiny islands are home to flocks of ospreys, white pelicans, and pink roseate spoonbills. The water in its ideal condition is translucent, nearly the color of the sky because it is clear as looking glass. At sunset the water is pink and orange.

"If you spent enough time here you'd see all the wonders of the sea," Mary Barley says.

It was among these wonders that George and Mary Barley planned to celebrate his fifty-eighth birthday in May 1992 doing what he loved best, tarpon fishing on Florida Bay.

George Barley was a wealthy real estate developer in Orlando, a seventh-generation Floridian and globe-trotting hunter and fisherman. He was a Republican and well connected through his involvement in a number of civic organizations, most recently as chair of the newly established Florida Keys National Marine Sanctuary Advisory Council. George knew President George H. W. Bush, who also enjoyed fishing on Florida Bay. George Barley was handsome and charismatic, with a full head of mahogany-colored hair, uneven smile, and dimpled chin. To describe him as energetic would be putting it lightly. Mary Barley assisted her husband with his business dealings.

For twenty years the couple had enjoyed long vacations in Islamorada, a lazy upper Keys village spanning six islands of beachy homes, lavish resorts, and eclectic shops and restaurants with names like the Twisted Shrimp. Locals consider the village the sport fishing capital of the world, where legends like Jimmy Albright, Cecil Keith, and Ted Williams honed their skills. George Barley liked fly-fishing for tarpon, an air-breathing beast of a fish capable of growing as large as 8 feet and more than 200 pounds. The couple would venture out early with a guide on a flats boat, motoring more than an hour to the Florida Bay backcountry where the tarpon would lurk in the shallows as they fed.

A flats boat is a midsize vessel with a platform on the back. After cutting the engine, the guide gets on top of the platform and uses a pole to maneuver the boat like a Venetian gondola. With the guide eyeing the looking-glass water from the platform, the Barleys would search the water from the front of the boat for a fin, ripple, flash of silvery tarpon. When someone spotted a fish the guide would stealthily pole the boat over.

"What you're trying to do is put a fly in front of the tarpon's mouth, maybe three or four feet in front of him so that he sees it and takes it," Mary says. "When you have dirty water you can't see them."

That was the problem on that day.

As a guide the Barleys had hired Hank Brown, an octogenarian who had fished and guided in Florida Bay for some forty years by the time I reached him by phone before his death in 2016. On the day of George's birthday fishing trip, though, Brown didn't remember fishing much.

Instead Brown motored the Barleys out near Sandy Key where a harmful algae bloom had turned the bay's lovely teal and purple water pea green and muddy. Hunting tarpon would be impossible here. You couldn't see even an inch into the slop.

"We were having these huge areas of die-offs. . . . There was just little or no life in them," Brown told me. "The clarity of the bay, it just turned into a sea of mud. [George], like myself, had never seen anything like this before, and it had a real impact on both him and Mary. . . . It was just like a tsunami coming in from the west, all this muddy water."

Since the late 1970s Hank Brown had watched as Florida Bay declined. Florida Bay is a lagoonal estuary, part of neither the Gulf of Mexico nor the Atlantic Ocean. The bay is an 850-square-mile watery outcropping between the two, hugging the peninsula so snugly that most of the bay lies within the boundaries of Everglades National Park. The bay draws little water from the gulf or ocean. The Keys bar the bay from the ocean, and the bay is so shallow that little water trickles in from the gulf except at Cape Sable, a shockingly white spit of beach in the verdant Everglades National Park accessible only by boat. Instead the bay gets most of its water from a river of grass to the north, the Everglades.

Seagrass carpets some 95 percent of Florida Bay's bottom, forming among the largest seagrass meadows in North America. The flats function as an ecological foundation for the bay. They serve as a nursery for economically important species like pink shrimp and are where sought-after reef fish like snapper forage.

In the 1970s anglers and guides began noticing subtle changes in Florida Bay. In the 1980s the Florida Keys Fishing Guides Association raised complaints about an experiment undertaken by the US Army Corps of Engineers and South Florida Water Management District aimed at providing better flood protection for farmers and residents of a remote part of South Florida known as the Frog Pond and Eight and a Half Square Mile Area. Under the experiment the water managers dropped the water level in canals next to Taylor Slough, the main course of freshwater into western Florida Bay.

When the seagrass began to die in 1987 the die-off spread quickly, destroying some 35,000 acres in four years and leaving behind a barren moonscape of a bottom beneath turbid water. At the same time, a massive harmful algae bloom flourished west of Florida Bay in the gulf, which anglers dubbed "the dead zone." Their diesel engines choked up when they ran their boats through the area, and they had to remove their traps and stop fishing there.

"This is a very large area, a few hundred square miles that grew and expanded toward Marathon at a mile and a half per day," George would write.

Making matters worse was a drought that left the bay even more parched for freshwater. Between 1970 and 1990 the bay got average rainfall or above during only four years. For many years there had been no hurricane that could have provided a good flushing. The bay's water grew noxious. It was hotter than normal and more saline with less oxygen.

In 1990 Everglades National Park experienced its worst fish kill ever in Florida Bay, in Garfield Bight and other coves. A year later the guides noticed the seagrass die-off had started up again and was racing through the western part of the bay. In less than a year some 65,000 more acres were lost, and another massive harmful algae bloom grew to be as large as 650 square miles. Scientists said the dead seagrass and stirred-up sediment from the denuded bottom produced nutrients that nourished the bloom. One scientist characterized Florida Bay's collapse as "unprecedented in history."

By the time of George Barley's birthday in May 1992 the problems stretched from Sandy Key in the northwest bay to Big Pine Key, 25 miles from Key West, moving through the middle keys' passes into Hawk Channel. Some $75 million in pink shrimp landings were lost, and the algae blooms killed nearly all of the sponges in the hardest-hit areas, another economic concern because juvenile lobsters and fin fish depended on the sponges for shelter from predators. Lobster landings also were down.

On the flats boat, the Barleys and Hank Brown stared into the pea soup water of George's beloved Florida Bay. It smelled of rot. George's anger rose.

"How could this happen?" he demanded.

2

THE BIG PICTURE

A little more than a century ago much of the Florida peninsula was underwater.

Often this information shocks people when I share it, and it is startling that, as the nation's third-most populous state, Florida is a frontier state, a subtropical one. Imagine Miami International Airport underwater, Palm Beach, Fort Lauderdale, and Miami itself as slivers of cities perched on the 5-mile-wide Atlantic Coastal Ridge, the Everglades' historic eastern bank. Today few places in Florida remain as they were then, although I have visited some—treasures near places like Walt Disney World, hidden well enough that even few Floridians have seen them.

The water begins as rain. Some 50 to 60 inches annually drench the peninsula, giving life to a watershed that, unlike any other on Earth, derives most of its water from rainfall. The watershed starts near Walt Disney World, in fact, with the headwaters of the Kissimmee River. Historically the river meandered 103 miles toward Lake Okeechobee. The state's largest lake, a shallow depression in the land, then spilled the water over its southern brim,

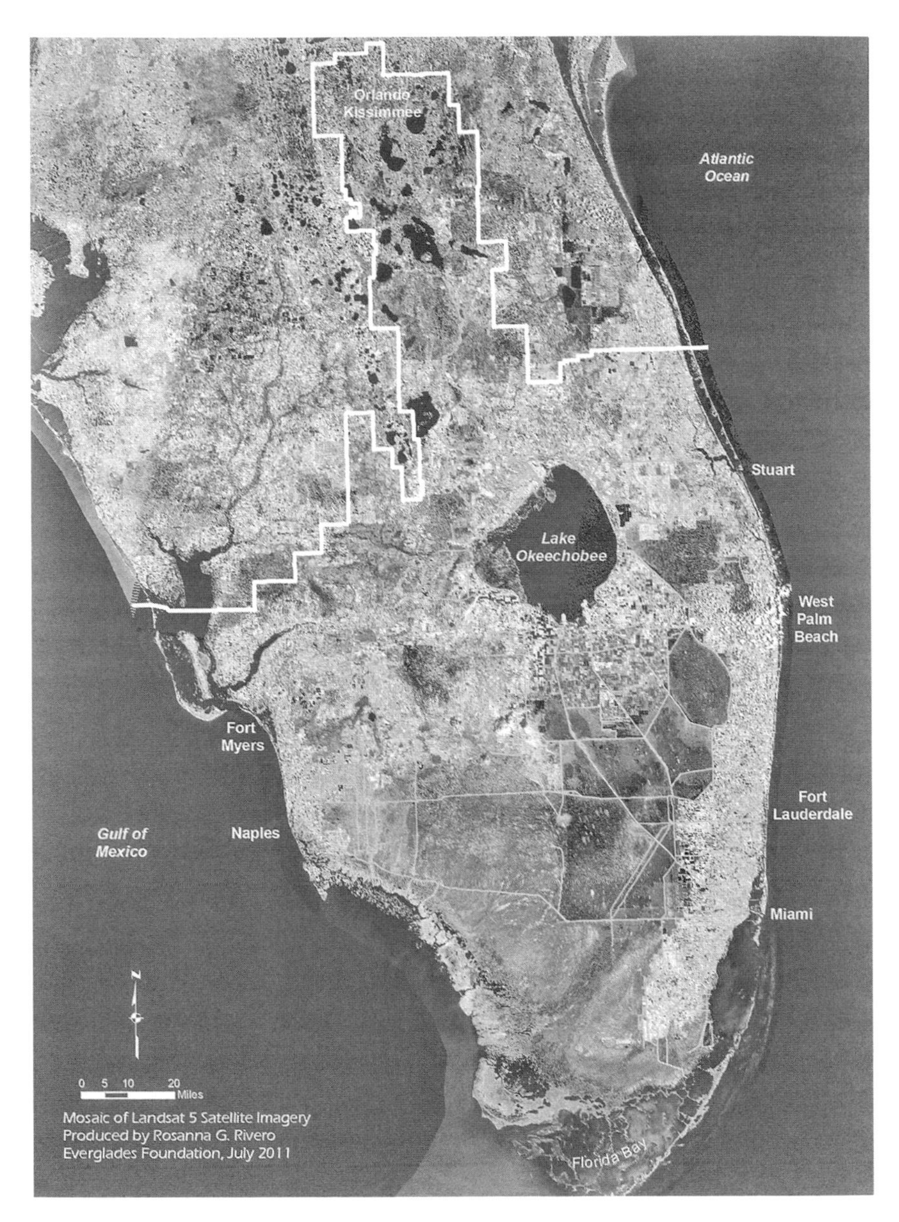

Everglades region

Courtesy of the Everglades Foundation

sending forth a shallow sheet spanning 70 miles at its widest, flowing slowly the length of South Florida in an arc toward Florida Bay and the Gulf of Mexico. This watery arc was the Everglades.

Long have the Everglades been elusive and mysterious, not a wonder of nature that easily commands our attention like the Grand Canyon or Mississippi River. The water courses 100 feet a day, above limestone bedrock that slopes two inches a mile. The watershed's shores were the first on the continent to be discovered by early European explorers. The region itself was among the last to be explored. The first studies of the watershed were completed only within recent decades. When Marjory Stoneman Douglas in 1947 published her classic book, *The Everglades: River of Grass*, coining the poetic phrase, the region still was not fully mapped.

"For four hundred years after the discovery they seemed more like a fantasy than a simple geographic and historic fact," she wrote. "Even the men who in the later years saw them more clearly could hardly make up their minds what the Everglades were or how they could be described, or what use could be made of them. They were mysterious then. They are mysterious still to everyone by whom their fundamental nature is not understood.

"Off and on for those four hundred years the region now called 'The Everglades' was described as a series of vast, miasmic swamps, poisonous lagoons, huge dismal marshes without outlet, a rotting, shallow inland sea, or labyrinths of dark trees hung and looped about with snakes and dropping mosses, malignant with tropical fevers and malarias, evil to the white man."

Hardly are the Everglades poisonous. The watershed teems with life, the water sustaining more than 1,000 plant and nearly 400 animal species. The watershed is home to more than 300 bird species, including egrets, herons, roseate spoonbills, and wood storks, and more than 65 reptile species, including alligators, snakes, and the American crocodile. There are 73 endangered or threatened species, including the Florida panther and West Indian manatee. The water flows through Big Cypress National Preserve, Biscayne National Park, Everglades National Park, John Pennekamp Coral Reef State Park, and Ten Thousand Islands National Wildlife Refuge.

The Everglades are vast, the nation's largest subtropical wilderness, spanning 6,000 square miles, and their flat terrain is important. Elevation

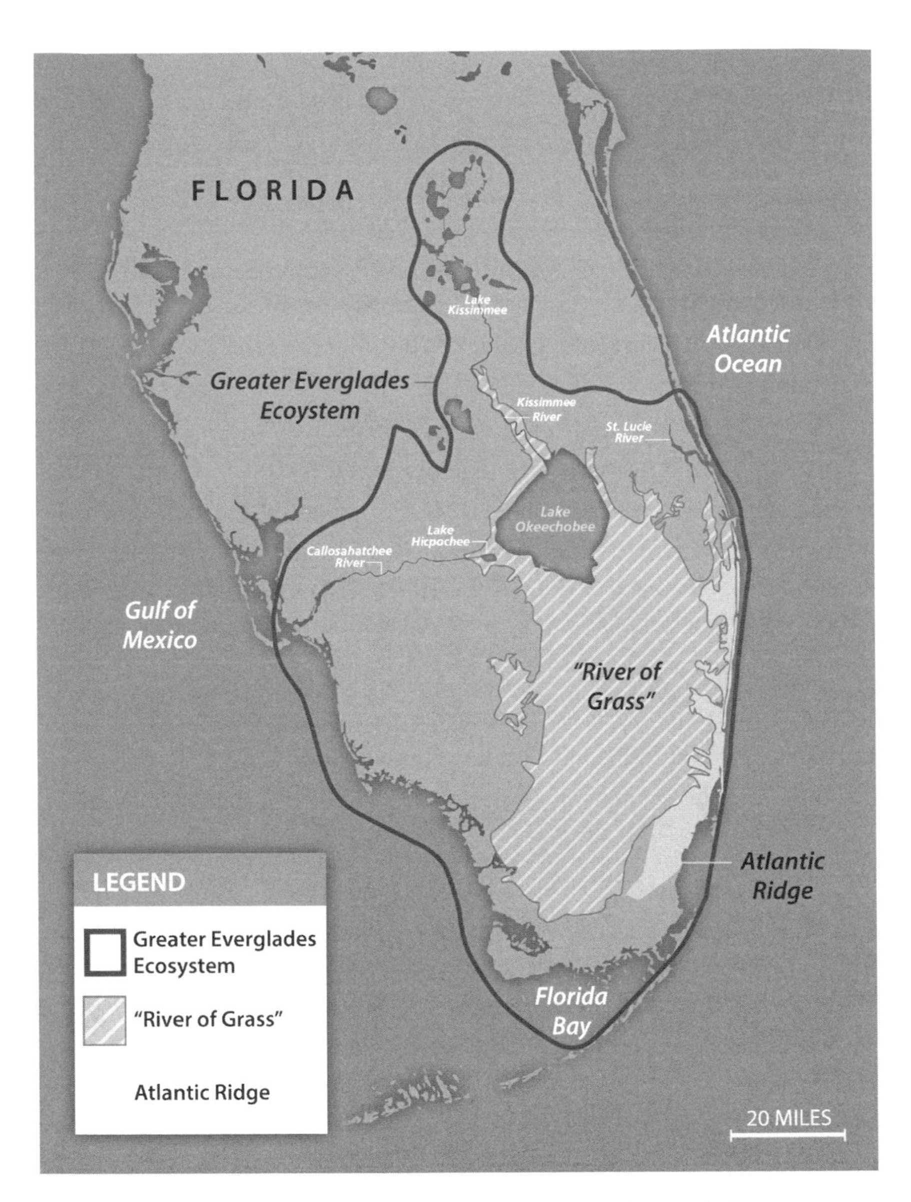

Everglades region. The crosshatched area shows the Everglades at their largest known historical extent.

Courtesy of the Everglades Foundation

Big Cypress National Preserve in 2008

Photo by Amy Green

variations of only inches are transformative. There are cypress swamps, lakes, marl prairies, pinelands, sawgrass marshes, sloughs, and tropical hammocks. What makes the Everglades a river of grass, though, is the sawgrass, which is not a grass at all but a sedge with sharp teeth jutting from the edges of each blade. The sawgrass marshes, best accessible by airboat, spread out green in every direction, an immense river of grass flowing imperceptibly beneath the blue dome of a sky that bestows the lifeblood rains. Pickerelweed joins the sawgrass in the watery sloughs. Islands of cypress trees loom dark in the distance. The sight of it all, the experience—I can tell you—is like running your hand over the irregularities of a piece of pottery wrought by the hand of God, uncomfortable and breathtaking.

"It's more about flat wide open spaces where you can really stretch your eyeballs," Mary mused one warm December afternoon as we strolled the paved trails of the Royal Palm Visitor Center within Everglades National Park. Tourists hardly noticed the small-statured woman in a blue woven sun hat and red sneakers, but for Mary it was a good day because the water was clear and there was plenty of it flowing in dark ribbons among the sawgrass. Happily she pointed at a baby alligator, a turtle, a sinewy white egret. "He's hunting," she said, watching the elegant crane.

IN 1900 SOUTH FLORIDA was a wilderness area, home to fewer than 50,000 people. After a series of varied attempts at draining the Everglades, Congress in 1948 authorized the Central and Southern Florida Project for Flood Control and Other Purposes, or C&SF Project, hailed at the time as the nation's largest civil works undertaking. To manage the effort the Florida legislature a year later established the Central and Southern Florida Flood Control District, the predecessor to today's South Florida Water Management District.

Some 8 million people today depend on the Everglades as a primary drinking water source. The population boom has left the river of grass half the size of the mighty watershed that ruled the peninsula more than a century ago. The Everglades are in pieces, the remnants held together by some of the world's most complex water management infrastructures, a network that has made modern Florida possible. More than 2,100 miles of canals,

2,000 miles of levees, 657 water control structures, and 77 major pump stations serve as life support for the river of grass, which is hardly a river anymore. The water courses in the wrong place at the wrong time and at the wrong rate.

No more does the Kissimmee River meander back and forth to Lake Okeechobee, although the ghosts of some of the meanders are still there. The meanders have been sliced through by a 56-mile canal, the C-38, carrying nutrient pollution from Orlando and cattle ranches to its south. The canal carries the pollution in a rush to the Everglades' watery historic heart, the Big O, as locals call it, without the benefit of the river's meanders and wetlands that served as natural filters. Today's restoration of the Kissimmee is aimed at resurrecting the cleansing meanders through the backfilling of some 22 miles of canal.

Lake Okeechobee has experienced some of the biggest alterations. No longer does the water spill over its southern brim. Canals have been excavated that draw the water east to the Atlantic Ocean and west to the Gulf of Mexico. Further canals divert water southeast to the Atlantic. Historically, freshwater was thought to be inexhaustible. The objective was to dispose of it as efficiently as possible through canals to the coasts. A dike across the lake's southern brim forever ended the water's historic flow, leaving the canals as the lake's only outlets. After a hurricane in 1928 caused flooding that killed thousands the dike was expanded to encircle all of the lake.

To the south the Everglades have been segregated further into the Everglades Agricultural Area, where sugarcane is the primary crop, and three water conservation areas, established later as the objective evolved from drainage to flood control. Palm Beach, Fort Lauderdale, and Miami have flourished on the east coast, the urban jungles like fraternal twins of the verdant Everglades. Everglades National Park is situated unfortunately at the peninsula's southern end, completely dependent on the region's water management infrastructure for water except for rainfall.

All of this alteration has resulted in the loss of some 1.7 billion gallons of freshwater daily to the Gulf of Mexico and Atlantic Ocean, leaving Florida Bay with a virtual trickle. By the early 1990s the bay was receiving less than one-tenth of its historic flow.

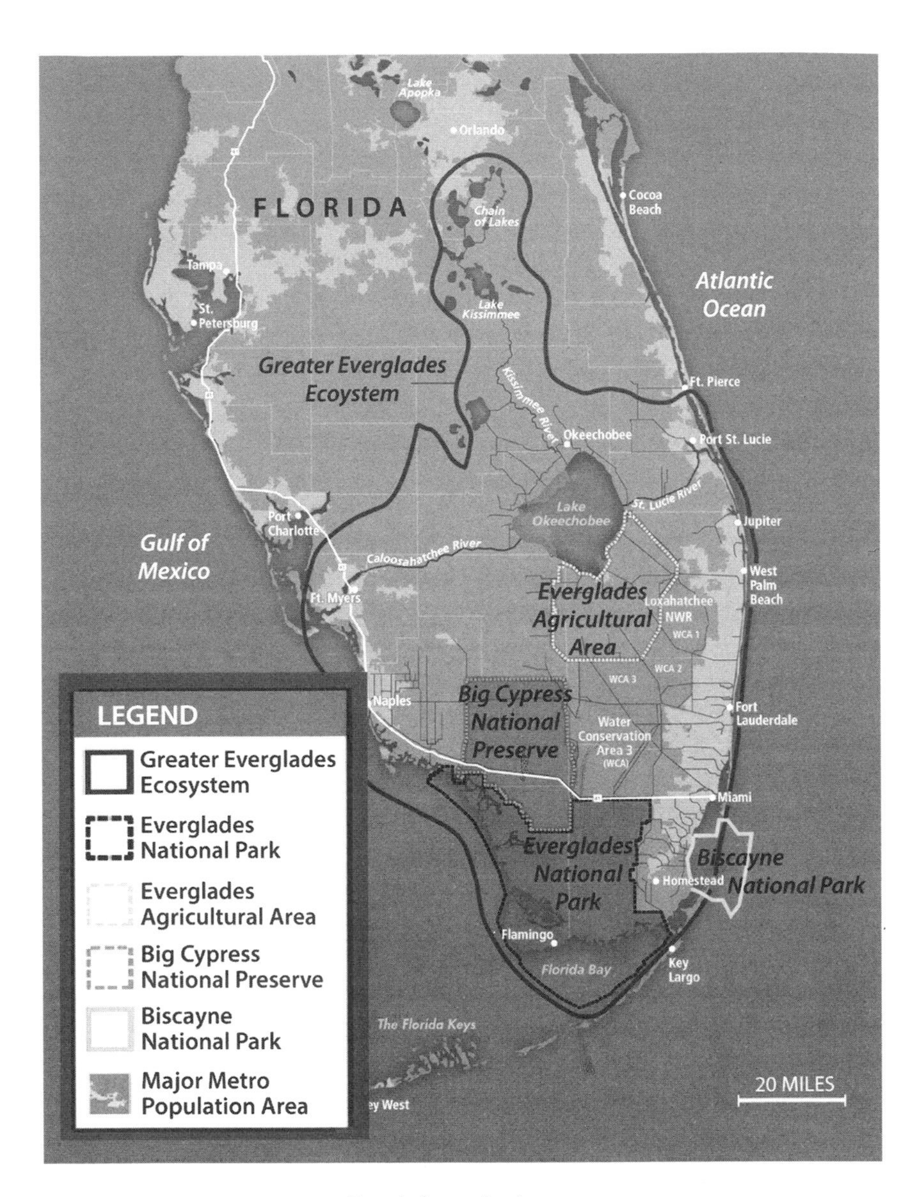

Everglades region in 2015

Courtesy of the Everglades Foundation

IN FEBRUARY 1992 the Florida Keys National Marine Sanctuary Advisory Council was formed, and at its first meeting the members called for action on Florida Bay from the Environmental Protection Agency and federal and state governments. The council oversaw what then was the nation's largest marine sanctuary, encompassing some 2,800 square miles from Biscayne Bay to the Dry Tortugas. Its chairman was George Barley.

George reached out to four prominent scientists and asked them to write letters summarizing their concerns about the ailing bay. He asked the scientists to address the letters to him as chair of the Florida Keys National Marine Sanctuary Advisory Council. Among the scientists George contacted was Ron Jones, director of Florida International University's Southeastern Environmental Research Center, whom he got to know over dinner at an Italian restaurant with a group of other scientists and environmentalists. Over expensive wine the group discussed a federal lawsuit under way against the state over sugar growers' pollution in the Everglades. The scientists and environmentalists wanted to pin the bay's problems on the same pollution, but Ron Jones disagreed.

He believed the bay's problems were more complex, that draining the Everglades had left the bay thirsting for freshwater and choking on salinity. George argued with Ron Jones—George was not used to being disagreed with—and after the meal Ron Jones believed he never would hear from George Barley again. Jones was wrong. Instead, the next morning he discovered an email waiting from George, one of a bevy of "Barleygrams," as the scientist called them, that he would receive in the coming years. In the message George thanked Ron for standing up to him and asked whether he could serve as a scientific advisor. A few months later a letter landed on George Barley's desk from Ron Jones:

> Let me begin by stating that the present [1992] condition of Florida Bay—algal blooms, seagrass die-off, turbid waters, hypersalinity, dead sponges, elevated temperatures and fish kills—doesn't bother me as much as the fact that Florida Bay has been in trouble for the better part of the last half of this century, and little attention has been paid to it until now. . . .

> For too long the Bay had seemed healthy to the casual observer, this was a major optical illusion! Florida Bay is supposed to be an estuarine environment, where the freshwaters of the Everglades meet the marine waters of the Gulf of Mexico. This type of environment has potential to be among the most productive in the world. Unfortunately for the Bay, South Florida is also an ideal environment for humans, and both the Bay and humans rely on freshwater for their survival. To the [detriment] of Florida Bay, human uses of freshwater don't mix well with natural system uses. An example is found in the effects of floods, life giving events for an environment like Florida Bay, life destroying events for human populations.
>
> Little thought was given to ecology when designing a water delivery system and control system to accommodate the ever growing population of South Florida. The result of this activity has been the slow conversion of Florida Bay to a hyper-saline coastal lagoon. Sure it looked good for many years, but looks can be deceiving. This strangulation of the bay is having a drastic effect on not only the bay itself but also the surrounding marine systems.

George Barley also reached out to Jay Zieman, who had studied Florida Bay for nearly three decades as an environmental sciences professor at the University of Virginia. Zieman wrote to George:

> While these high salinities are exacerbated by recent climatic anomalies that my research group at the University of Virginia has documented to be currently occurring in south Florida, their root is in the human curtailment of the historic waterflow into Florida Bay from the Everglades. It is difficult to find a problem in Florida Bay that is not traceable, in whole or in part to this lack of fresh water. It is the key to everything.
>
> The seagrass dieoff story also highlights other problems that are strangling the research efforts within the park. Within Florida Bay seagrasses cover 95 percent of the bottom. They

> are the keystone around which all of the other species such as pink shrimp, sea trout, gray snapper, redfish and many others depend. Without the seagrasses there would be just a barren mudflat and dirty, turbid water, as many fishermen are finding now. The response to the potential collapse of this ecosystem has been abysmal. Because they are below water they are out of most people's sight and therefore out of mind. This includes the minds of many officials who should be not just concerned but quite frankly scared at this latest surge in dieoff. Because they are not brilliantly covered they do not get the press attention that coral reefs get.
>
> South Florida in general does not recognize how dependent it is on seagrass beds. You as a fisherman are well aware of the close link with bonefish, sea trout, redfish, snook, and many other food and gamefish that are totally dependent on this ecosystem. However it is also the main nursery of the pink shrimp in the area. As seagrass habitat is degraded, the commercial pink fish fishery will follow. What few people recognize, however, is that many of the coral reef fish are dependent on this ecosystem. Many of the schools of grunts and snappers that are found on the reef school there by day for protection, but at night spread out and forage in the seagrasses behind the reef for food. If seagrass dieoff does not abate, the economic potential for the region is grim.

Zieman wrote in later correspondence that Florida Bay was not dead, as many feared. Far from it. The phytoplankton blooms and microbial decomposition of seagrass indicated the bay was vividly alive. Nevertheless, the bay was changing, transforming from an ecosystem with "high habitat values," as the scientist put it, to a much less desirable one with "far lower habitat values."

"This new system is now fighting for its life," he wrote.

George fired off the letters to government agencies, lawmakers, the media—pretty much anyone who was relevant. At one point he wrote to Hunter S. Thompson and invited the gonzo journalist to fish with him in

Florida Bay so that he could see the problems for himself—and write about them. Meanwhile the Florida Keys National Marine Sanctuary Advisory Council demanded action from Interior Secretary Manuel Lujan Jr. and Teresa Gorman, environmental aide to President George Bush. The deputy director of the National Oceanic and Atmospheric Administration (NOAA) convened an interagency meeting, where it was determined that the problems were already being handled by NOAA's National Marine Fisheries Service and other agencies.

George Barley did not think the agencies were doing anything! He began devoting most of his personal time to the cause. He borrowed a plane from a wealthy friend and—a half-dozen passengers at a time—flew a thousand people over the bay so they could see what was happening.

"What people saw stunned them," George told a newspaper.

In August 1992 George helped arrange a helicopter tour for Governor Lawton Chiles, who would describe the bay as in the midst of an environmental crisis. By now the harmful algae blooms had the appearance from the air of paint flung on a canvas—dark blue-green here, a stream of brown there, a puddle of yellow over there.

"What the governor saw today was an environmental collapse unpreceded in Florida history," George Barley said. "It's a national catastrophe."

Indeed, the governor said what he saw startled him.

"I remember the waters being much clearer," Chiles said. "You can see some major changes out there. The grass beds are dying, and there's an algae bloom that you can see going all the way to the horizon. It's already having an effect on the fishing industry. . . . And the way it's going right now, it's going to go out to the coral reefs. If we kill that, that's our tourist industry."

A day later the Florida Keys National Marine Sanctuary Advisory Council met with the governor. Among other things the council called on Chiles to order that growth management plans in South Florida protect Florida Bay and the Everglades and to direct the South Florida Water Management District to prioritize restoring water flow into Everglades National Park.

The council and scientists presented their concerns before the South Florida Water Management District as well as an interagency group appointed by the Florida Department of Environmental Regulation, as it was called at the time. In February 1993 Interior Secretary Bruce Babbitt announced at a meeting of the Everglades Coalition, an alliance of organizations dedicated to the watershed, that he would establish an interagency committee to work on Florida Bay and Everglades restoration. The water management district held two public meetings in the Florida Keys where some seven hundred citizens clamored for action. That July, three congressional subcommittees met in the Keys to hold field hearings on Florida Bay.

One Associated Press article characterized Florida Bay's problems as the worst crisis the Keys had faced since the Cuban missile crisis. George Barley was just getting started. He believed that once government leaders understood the problems they would intervene, Mary recalls of her husband. Why not? Florida Bay was an important natural resource, supporting the Keys' $600 million fishing and tourism industries.

"The mistake we made was we thought science would rule the game," she says. "What we found out was science and fact are different from political science. It's not that we were naive. I think George really thought they were more honorable than they were. None of them were looking at the big picture. . . . They didn't care about how is the water going to get down into Everglades National Park or down to Florida Bay. [They were] just thinking right here in this little block, and so everything in the system was done like that from Orlando down. . . .

"If George did anything, what he did was he gave everybody the big picture, and nobody's ever turned back."

3

BIG SUGAR

In July 1993 George Barley peered through the window of the six-seater plane at the cane fields below, 520,000 acres of them unfurling in squares of green and brown knitted with canals carrying water south to the Everglades. Above the propeller's din Ron Jones talked as George peppered him with a zillion questions. The fields' polluted water was one problem, Jones said. There also were the water guzzlers of Miami, Fort Lauderdale, and Palm Beach, the water management infrastructure sending the water in all the wrong ways.

"For me it was probably the best airplane trip I've ever been on," recalls Jones, who is prone to air sickness. "It was the first time I had ever flown the whole area, and it put together in my mind the whole problem and what were the various angles in the problem and how are they involved, and I think George had the same experience."

The plane had taken off near Miami and soared west over the river of grass and north toward Lake Okeechobee. George always liked chartering flights to show people around, and he and Ron Jones asked the pilot to land for lunch in Florida's Everglades Agricultural Area.

SUGARCANE ARRIVED IN FLORIDA in about 1565 when Spanish colonial governor Pedro Menéndez de Avilés brought the crop to the St. Augustine area, although commercial production did not begin until 1767 in the New Smyrna colony and was disrupted by the American Revolution. By the 1800s unsuitable soil and inclement weather had foiled new attempts at commercial operations in Central Florida, forcing the pioneer growers in the late 1800s and early 1900s to move south.

The growers found abundant water and sunshine on the southeastern shore of Lake Okeechobee and finally settled there in the 1920s. During the following decades several events led the industry to flourish: namely, the development of varieties suitable for the region's semitropical climate and of the infrastructure necessary to keep the region dry during the rainy season. Then in 1960 the United States stopped importing sugar from Cuba. Before that three South Florida mills produced up to 175,000 tons of raw sugar annually from some 50,000 acres. Afterward those numbers ballooned to eleven mills churning out 572,000 tons annually from 223,000 acres.

In 1948 a World War II veteran named Ray Roth who had grown up on a successful but small vegetable farm in Ohio moved to Belle Glade, Florida, to pursue his own dream of farming. For several years he toiled on leased land until by 1955 he had saved enough money for land of his own. He planted vegetables and in 1962 bought more land and added sugarcane. The farm continued to thrive, and after Ray's death his son Rick took over. By the time I met Rick Roth in 2019 at one of his cane fields over July 4th weekend, Roth Farms spanned some 4,000 acres in five locations in the Belle Glade area, producing some $10 million annually in produce and sugarcane, making it a midsize farm in the Everglades Agricultural Area.

"My father was a very good farmer. He was farming from when he was like two years old," Rick told me. He wore a bright-blue T-shirt reading "Sweet Corn Fiesta," gold chain necklace, and navy shorts. He was charismatic, with a sweep of silver hair, and spoke with a baritone drawl. By now he was semi-retired and a Republican state legislator, and his son Ryan Roth mostly ran the farm.

"We grew rice every summer in rotation with vegetables and sugar," Rick continued. "I grew sod after my dad passed away in 1986, and so we were growing vegetables, sugar, rice, and sod from '87 to 2009, and then we got out of the sod business because of the financial crisis. If you don't build houses you don't sell sod. We still grow radishes, leafy vegetables, lettuce, about 20 different vegetable crops—celery, sweet corn, green beans. We're very diversified, but growing sugar and vegetables in rotation is the perfect way to farm. It does so much good for the land."

The Everglades Agricultural Area is the nation's largest producer of sugarcane, raising half of the country's crop on a 700,000-acre swath representing 27 percent of what used to be the Everglades. Here the farmers of communities like Belle Glade, Clewiston, and Pahokee cultivate citrus, rice, vegetables, and some 2 million tons of cane annually, constituting a quarter of all of the sugar—cane and beet—produced in the United States. The two largest companies are U.S. Sugar Corp. and Florida Crystals Corp., each representing more than 40 percent of the acreage. Midsize and smaller farms like Roth's make up the rest. The Everglades Agricultural Area is so compact, so removed from the state's beaches and urban areas, that many Floridians have never seen a cane field.

"Because we farm so many months out of the year and so much land, I think this is the most significant single area for growing food in the United States," Roth says. "California as a state grows a lot more food, but they grow it over a 400-mile area. We're talking about farming in one area year-round. Nobody else does that."

Sugarcane is a tropical grass native to Asia, where it has been a popular part of gardens for more than four thousand years. Methods for manufacturing sugar from the plant were developed in India in about 400 BC. Christopher Columbus brought the plant to the West Indies in the sixteenth century, and today sugarcane is cultivated in tropical and subtropical regions across the globe. Some 75 percent of the world's sugar comes from sugarcane.

In the Everglades Agricultural Area locals describe the soil as black gold. Belle Glade welcomes visitors with a roadway sign proclaiming, "Her Soil Is Her Fortune." The river of grass enriched the muck soil with nitrogen, a nutrient found in fertilizer.

"It's black. It's a very light soil. It's made out of decayed plant material. You think of it like a river bottom or lake bottom or areas that have stayed wet for a long time," Rick Roth says. "This is decomposed vegetative material that was formed over thousands and thousands and thousands of years by Lake Okeechobee overflowing its boundaries."

From this fertile blackness sprouts cane up to 15 feet tall, grassy and jointed like bamboo. Crack it open, and it tastes like sugar. Roth says cane here actually is easier and can be more profitable to grow than vegetables because the cane requires less fertilizer and is less sensitive to weather events like freezes or floods. From October to May the cane is harvested day and night by mechanical harvesters, although before the early 1990s workers, many of them low-wage migrants from the Caribbean, harvested the cane using machetes beneath the brutal Florida sun. The backbreaking work and the growers' treatment of the workers prompted litigation and a federal investigation and for years was a source of deep controversy and shame for the industry, leading to the switch to mechanical harvesters.

The harvest process begins with burning the fields to clear them of dead leaves and other debris. From a distance it appears as though the Earth itself is exhaling after a long drag. The smoke can cling to the communities of the Everglades Agricultural Area, dusting them with burnt bits of plant material that locals describe as snow. This, too, has been a source of controversy and in 2019 prompted a class-action lawsuit led by the Sierra Club, as some locals have complained of respiratory symptoms, though the American Lung Association and Florida Department of Health have concluded the air quality meets codes.

The mechanical harvesters are truck-like machines that cut the cane and dump it into the wagons of tractors, which can haul multiple tons to highway trailers or rail cars for transport to the mills. The largest mill belongs to U.S. Sugar, which alone processes some 42,000 tons of cane daily, grinding it all at one of the largest mills of its kind in the world. Up to 9 percent of the country's sugar annually comes from the company, and its mill towers above Clewiston and Lake Okeechobee's southern shore with a metaphorical might matching that of the expansive Big O.

At the mills the cane is crushed between heavy rollers, its juice squeezed out. The juice is processed into a sucrose solution that is then concentrated as the water is boiled off. Raw sugar crystals form, and the final traces of molasses are removed at refineries located from South Florida to New Orleans to

U.S. Sugar Corp. in Clewiston
Photo by Amy Green

Yonkers, New York. What makes U.S. Sugar's mill so large is that it combines with a refinery into one. The refineries produce the granulated sugar we buy in grocery stores and also liquid sugar, packaging all of it in consumer-sized bags or shipping it in tanker trucks for use in the baked goods, beverages, and processed foods we enjoy.

U.S. Sugar was founded by Charles Stewart Mott, the automotive pioneer of General Motors, and the foundation bearing his name now cites the environment and clean freshwater as two of its causes. The company owns an independent short-line railroad, the South Central Florida Express, and is the largest employer in Clewiston, which bills itself as "America's Sweetest Town." The main road through here is Sugarland Highway, and the high school stadium is Cane Field. The town's other large employers include the local government, large farms, and banks.

The Everglades Agricultural Area has experienced times of great economic hardship. During the 2009 recession unemployment and poverty rates spiked as high as 45 percent in Belle Glade and Pahokee, according to the US Census Bureau. When I visited in 2012, conditions in some neighborhoods were like those of Third World countries, with whole families living in seemingly unhabitable dwellings. Efforts to diversify the economy have failed.

IN 1988 US ATTORNEY DEXTER LEHTINEN, the federal government's top prosecutor in South Florida, filed a lawsuit against the state. Specifically, the suit asserted that the South Florida Water Management District and the state's Department of Environmental Regulation (as it was called at the time) had violated their own water quality laws by allowing sugar growers in the Everglades Agricultural Area to pollute Everglades National Park and the Arthur R. Marshall Loxahatchee National Wildlife Refuge. The lawsuit did not name sugar growers as defendants, but after previous Everglades restoration efforts had set modest goals the legal action represented a first effort aimed directly at the pollution of the very politically powerful sugar growers, and the growers fought it bitterly.

George Barley closely followed the action. In 1990 he took it upon himself to write to Lehtinen. George was especially interested in how much sugar growers might pay to clean up the pollution:

> Making note of the recent proposed acquisition from the sugar cane farmers of 70,000 acres to be paid ⅓ by all the tax payers of the State, an additional ⅓ by those tax payers in the South Florida Water Management District, and the balance of ⅓ to be paid in a tax deductible form, over a period of ten years or so, by the sugar cane farmers (I have not bothered to calculate their true net, discounted cash cost for this) I am appalled at the thought that we are being called upon to correct the pollution they are causing. I hope you will continue to prosecute your lawsuit vigorously for some more equitable resolution.
>
> Is it possible the sugar cane farmers can be sued for damages? I might know a few private citizens with the resources to pursue such a course of action if you thought it would have merit. I presume the federal government would have pursued

> this if it had merit but am, of course, not fully familiar with the complexities of this case.
>
> Please keep up the good job you are doing.

A few months later George wrote to Assistant Attorney General Richard Stewart at the US Department of Justice, after an *Orlando Sentinel* article implied the litigation was being delayed to protect Governor Bob Martinez's reelection bid. Martinez would lose to Lawton Chiles, an attorney who campaigned on a pledge to quickly find a settlement to the lawsuit. George wrote:

> Governor Martinez and the State and National Republican parties have accepted important financial support from sugar cane farmers and if the SENTINEL's allegations are true it would be a scandal of the first order.
>
> The proposed settlement of this issue by Governor Martinez in which the sugar cane farmers would be given indemnity from further payment after agreeing to presumably a tax deductible 10 year payment of only $20 million (with the tax payers to take up the balance) to buy land from these sugar cane farmers and construct improvements to clean up their pollution is a pitifully inadequate settlement. I seek your assurance the Justice Department is not delaying this case for political or any other reason and will move ahead vigorously.

By 1991 the lawsuit was the most complex and highest-profile environmental lawsuit in the nation. The state spent millions of dollars defending the suit, and it wasn't long before court clerks had to wheel the case file around in a grocery cart. In a Miami courtroom in May Lehtinen held up a glass of water and said: "We sued over that water because it is dirty water. We will be only satisfied when all of the plans have an effect on the water that's in that glass."

The federal attorney went on. If the state was so eager to clean up the water, he wondered, then why wouldn't the state attorneys "stand up at this podium and say that this water is dirty?"

Governor Lawton Chiles spoke next.

"I am ready to stipulate today that the water is dirty!" he blurted. "I am here, and I brought my sword. I want to find out who I can give that sword to!"

Across the courtroom jaws dropped. Clearly the state's own attorneys were not prepared for the governor's admission, although Chiles was stating the obvious: that the water flowing from the cane fields was harming the Everglades.

"What I am asking is to let us use our troops to clean up the battlefield now, to make this water clean," the governor continued. "We want to surrender!"

In July 1991 the federal and state governments reached a settlement acknowledging two main threats to the river of grass: the watershed was not getting enough freshwater, and the water it was getting was polluted by sugar growers. State scientists admitted that water flowing into Everglades National Park had not met state standards during eight of the past ten years, and that in the Loxahatchee wildlife refuge phosphorus amounts were twenty times natural levels. Seven months later a federal judge signed a thirty-six-page consent decree ordering the state and sugar growers to begin implementing an eleven-year cleanup plan. The growers opposed the agreement and followed up with legal challenges that by 1993 continued to tie up the issue in court.

THE PLANE LANDED, AND GEORGE BARLEY and Ron Jones had lunch at a little diner among the cane fields.

Ron Jones was US Attorney Dexter Lehtinen's star witness against the state. At the heart of the litigation was phosphorus. Not only did sugar growers' location south of Lake Okeechobee interfere with the river of grass's historic flow. Their fertilizers poured phosphorus into the river of grass where swamp forest and sawgrass once had cleansed the water of the nutrient, leaving the Everglades with only scant nutrients to the south.

Instead phosphorus levels were 120 times the natural amount. Ron Jones's research proved that when phosphorus levels exceeded 10 parts per billion, or 10 cubes of sugar dropped in an Olympic-sized swimming pool, oxygen began to vanish from the water and algae flourished. Sawgrass sprouted abnormally tall, and cattails grew so thick birds and fish couldn't land or swim in them, much less feed. Jones also demonstrated that the spread of phosphorus in the river of grass followed the same path as that of the water flowing from the Everglades Agricultural Area.

The growers, led by the large ones like U.S. Sugar and the Florida Sugar Cane League, disputed and attacked Ron Jones's findings. They blamed state mismanagement of South Florida's water for the Everglades' problems and in court filings actually accused the federal government of being in "complicity" with the same state leaders named in the lawsuit, with the goal of driving growers out of business and the region where some had raised cane for generations. Marjory Stoneman Douglas, author of *The Everglades: River of Grass*, had written to the governor at one point to express her "violent conviction" that the region ought to be reflooded.

The 1991 settlement would initiate one of the largest scientific experiments ever undertaken at the time. It called for a 32,600-acre marsh that would filter the water of phosphorus as the Everglades once had. The state agreed to reduce phosphorus levels by 80 percent in the next six years and even more by 2002. If it did not look as though the filter marsh would work the state was required to add more acreage to the marsh. The state also had to make sure the filter marsh would not reduce the amount of water flowing into Everglades National Park or the Loxahatchee wildlife refuge. Growers also would have to stem the phosphorus flowing from their fields by 25 percent.

For Ron Jones the litigation was difficult, and he considered George Barley a dear friend.

"George was very special to me," Jones told me.

Over lunch the two talked of their flight.

"He was full of ideas of who we had to show this to next," Jones says.

On the way back they asked the pilot to fly over the Florida Keys and Florida Keys National Marine Sanctuary. Then they landed near Miami.

"He was a person, George, who was willing to spend his own money to understand the situation that had absolutely nothing to do with his business," Jones says. "He wasn't making anything off of this. He was losing off of this, obviously."

IN JULY 1993, the same month George Barley and Ron Jones flew to the Everglades Agricultural Area, President Bill Clinton's administration unveiled a $465 million plan to end all of the litigation over the Everglades and sugar growers' water. Flanked on the stage of an ornate Department of the Interior auditorium by state leaders and sugar executives, Interior Secretary Bruce Babbitt described the plan as "the largest, most ambitious ecosystem restoration ever undertaken in this country."

"Of some 350 national parks," he said, "I don't think there's any question that the most imperiled has been the Everglades."

Under the agreement, reached after intense late-night negotiations, growers would pay as much as $322 million over twenty years to clean up the water. The federal government, the state, and South Florida water users would contribute the rest. In its news release, the Department of the Interior stated, "Under the agreement, the largest burden for payment will be shouldered by south Florida's agriculture industry." The news release went on to quote Babbitt: "Sugar industry leaders negotiated in good faith and have accepted a lion's share of the financial responsibility."

"Today the Clinton administration delivers," Palm Beach sugar magnate Alfonso Fanjul declared.

The money would fund the purchase of some 40,000 acres of farmland to be converted into six artificial marshes that would filter the water of phosphorus. Parts of the region's water management infrastructure would be retooled to direct more freshwater toward the Everglades. Growers also agreed to stem their flow of phosphorus by 25 percent within two years and gradually thereafter through 2013. Doing so at a faster pace would cut their share of the cost by as much as $90 million. Further details were to be worked out during the next ninety days.

The environmental groups were incensed. Not only was Babbitt standing there shoulder to shoulder with Fanjul, but to them the agreement was vague and hardly held the growers accountable. Cleanup deadlines were extended, and they believed taxpayers bore most of the cost. The news conference ended abruptly when Joe Browder voiced his disapproval.

"It's an absolute betrayal, Bruce, and it won't stand," said Browder, a well-known advocate for Florida's environment who had influenced environmental policy at the Interior Department under President Jimmy Carter.

Attorney Dexter Lehtinen, now representing the Miccosukee Indian tribe, said the agreement would delay the Everglades cleanup. George felt the agreement overvalued growers' contribution because it failed to consider inflation or interest. He pointed out that $322 million paid over twenty years really was worth $153 million in 1993 dollars.

"I'd characterize it as a sellout," he fumed to one newspaper. "Has the government really agreed that sugar can't be made to pay any more?"

Weeks later he wrote an op-ed for the *Orlando Sentinel*:

> I grew up in Florida, and for the last several decades I have listened to politicians and bureaucrats periodically proclaim "victory" over the restoration of the Everglades. Meanwhile the fragile ecosystem becomes sicker every year.
>
> Why should Central Floridians care about the Everglades?
>
> The Everglades is one of the largest and most diverse wetland and aquatic systems in the world. Supporting a vast amount of wildlife and sea life, it is an essential source of clean water for Florida.
>
> It includes the Kissimmee River, Lake Okeechobee, the river of grass, Florida Bay and the Florida Keys. The jobs, tourism and other economic activities it supports are major generators of tax revenue for the state of Florida, and all our taxes will go up as it declines.
>
> For all of us who live in Florida, the Everglades system is our Yellowstone, our Grand Canyon—our heritage. If we do not stand up for the Everglades now, it will die.
>
> And that is what's happening: Despite the reassurance of the politicians and bureaucrats to the contrary, the Everglades is imperiled. Some 100,000 acres of sea grasses have died in Florida Bay, which have spawned up to 600 square miles of algae blooms that are threatening the coral reefs—and the economy of South Florida. Meanwhile, the U.S. Park Service calls the Everglades its "most endangered" national park.
>
> This is a tragedy and a disgrace we must address.

The latest proclamation of "victory" occurred in Washington last month. U.S. Interior Secretary Bruce Babbitt and Alfonso Fanjul, a Palm Beach sugar baron, stood on a stage and announced a grand, new partnership to "save" the Everglades. "Save" the Everglades from what?

"Save" it from the very pollution generated by Fanjul and other powerful sugar growers who influence Florida and national politics because of their connections and largess.

But there was one "partner" absent from the stage: the Florida taxpayer, you and me. Why? Because the government has decided we should pay to clean up sugar industry's pollution. Perhaps that's what Fanjul meant when he proclaimed on stage, "The Clinton administration delivers."

Now guess who was a major supporter of presidential candidate Bill Clinton? Yes, Alfonso Fanjul. Want another surprise? His brother, Pepe, was a major supporter of President Bush. Get the picture? Do you see what the Clinton administration "delivered"?

The current cleanup proposal will use between $3 and $4 of public money for every $1 of sugar money, with taxes to be raised from Orlando to Fort Myers to Key West. And even then, the settlement doesn't really clean up the Everglades. Sugar is still permitted to pollute—just not as much. Moreover, the sugar barons will get to use public lands to create many of their water-cleansing ponds. The taxpayers will again pick up the tab. And sugar has 20 years to pay up, with no interest (we'll pay that, too).

Charles Keating and the S&L crisis taught us a painful lesson about how government deals with powerful interest groups. Now, Florida taxpayers are about to learn a new lesson.

Let me elaborate.

The South Florida Water Management District is a group of bureaucrats supervised by a board of appointed commissioners. The district failed to police the destruction of the Everglades, which the sugar industry has been polluting for decades.

Two years ago, the U.S. government sued the state of Florida, demanding an Everglades clean-up. Eventually, the district admitted to the obvious and agreed to the federal charges.

A settlement agreement was drawn up. The sugar barons launched a heavyweight legal assault that tied that agreement in knots.

And that brings us up to the meeting last month. As I said, the politicians and bureaucrats and sugar were all on stage. The news media were summoned. The great "partnership" to clean up the Everglades was announced.

A star line-up of political figures, from Secretary Babbitt and Sen. Bob Graham to Gov. Lawton Chiles and Lt. Gov. Buddy MacKay have endorsed this partnership.

The politicians have decided, in this case, the polluter doesn't have to pay up. You and I do.

It's not for lack of funds. The sugar industry is very, very rich and can well afford to pay for its mess in the Everglades. That would leave taxpayers' money free for the real job of restoring the Everglades and Florida Bay. Now, where will the money come for undoing what the Corps of Engineers has done to the river of grass and Florida Bay?

Is the survival of the Everglades worth a second look? Should we not demand a real settlement?

Long after we are gone, the Everglades ecosystem will be our legacy—to our children and the rest of the nation. We must summon up the courage and political will to re-examine the "partnership."

So what is a real solution? Four key points:

Scientifically approved stormwater treatment areas must be built on sugar's land, and they must stop polluting the Everglades.

Sugar must pay their fair share of the capital cost. They are probably paying around 20 percent of their share in the proposed settlement—the taxpayers are paying the rest—some $200 million plus, according to a National Audubon analysis.

The first phase of the cleanup must be in place as close as practicable to the date stipulated in the court settlement, 1997. The proposed agreement puts this off until 2004.

The agreement should provide for clean water by the year 2002, as legally mandated.

Can this matter be settled? Yes, but only if the Justice Department under Janet Reno and Florida under Gov. Chiles refuse to buckle under and bring the full force of federal and state authority to bear to make them clean up their pollution.

4

THE POLITICS OF WATER

Beneath a bright blue sky of wispy clouds I stepped on a boat in July 2018 and ventured into the hydrological heart of the Everglades.

Before leaving the dock of Roland and Mary Ann Martin's marina in Clewiston I could see the toxic algae, like streaks of fluorescent green paint in Lake Okeechobee's dark water, but it wasn't enough to smell strongly. By the time Ramon Iglesias, the marina's manager, had motored me to Coot Bay off the lake's southern shore the algae seemed like an afterthought. The sun shone happily as Iglesias boasted of the lake's bounty: tournament-sized bass weighing 7 pounds, some 8. Bluegill, crappie, tilapia. I couldn't help but relax as a breeze cut through the afternoon heat. Then I turned to peer over the boat's side. In the water I could see what looked like millions of tiny air bubbles—green, the color of toxic algae.

"I'm not denying we have algae on the lake," said Iglesias, who was born and raised in Clewiston and for fourteen years had managed the marina established by the famous fisherman Roland Martin. The marina now was owned by Martin's ex-wife Mary Ann. "Unfortunately what they're showing . . . as far as the media is concerned, our customers, they really think it's 90 percent covered with a bright sheen of algae that is guacamole-thick, and that is not the case."

He was right. The worst blooms Floridians were hearing about, thick as guacamole and smelling like excrement, were not here in Coot Bay. Instead those blooms mostly were in the basins and canals of Lake Okeechobee and the St. Lucie and Caloosahatchee Rivers, pockets that lacked strong currents that could break up pileups of layer upon foul layer, as I would observe later that afternoon at the Moore Haven Lock and Dam, where the lake flows into the Caloosahatchee. In pockets such as those the smell could be as strong as excrement—times a million.

It also was true that as much as 90 percent of the 730-square-mile Lake Okeechobee, along with significant parts of the St. Lucie and Caloosahatchee Rivers, were sullied to some degree that summer with blue-green algae, or cyanobacteria, as shown in satellite images. Samples of the algae tested positive for microcystin, a potent liver toxin. The outbreak coincided with one of another toxic algae species called red tide, or Karenia brevis, that began in southwest Florida in the Gulf of Mexico and eventually gripped much of the peninsula. Red tide exposure can cause respiratory symptoms like coughing, sneezing, and watery eyes. The algae was a disaster in Florida, especially for anyone who derived their livelihoods from these waters, such as Iglesias. I asked him whether he would eat fish hooked in Lake Okeechobee.

"Yeah, I would," he said, his mild southern accent apparent above the boat motor's din. "All day long."

I felt for him, but I would not have eaten the fish.

THE DISASTER WAS WROUGHT NOT BY GOD but humankind, a toxic side effect of one of the most complex water management systems in the world. In the river of grass the water begins as rain, but its natural course ends there.

Today elaborate water management infrastructure makes modern life possible for some 8 million people where the river of grass once flowed. Running the machine are the South Florida Water Management District and US Army Corps of Engineers, which hold in their grip the Everglades' lifeblood water. Within the army corps are ten men and women in particular who direct the flow from Lake Okeechobee into the St. Lucie and Caloosahatchee Rivers while basically answering to no one—not agricultural, environmental, or government leaders. It is a harrowing balancing act in a delicate watershed where inches matter.

"They are completely omnipotent," says Paul Gray, an Audubon Florida scientist who for more than three decades has studied Lake Okeechobee. "They make decisions to open and close the gates, and no one can contradict them."

At the heart of their decisions is a half-inch-thick document bureaucratically named the Lake Okeechobee Regulation Schedule, developed after Hurricane Katrina through a public process. In 2019 the army corps began a three-year effort to update the rules for managing the state's largest lake as part of a nearly $2 billion project to refurbish the lake's aging dike. Included within the document's pages is a decision tree, basically, beginning with the lake's water level. When for instance the lake is too high, as it was that toxic summer after record rain in May and Hurricane Irma the year before, arrows point through a series of scenarios considering factors such as precipitation, time of year, and tributary levels: Do forecasts call for a lot of rain or a little? Has hurricane season just begun or ended? Are tributary flows heavy or light? At the end of this succession of arrows lay the answer to the question vexing everyone that summer: What are we going to do with all of this water?

FEW OF AMERICA'S WATERWAYS flow freely anymore, though not many are subject to water management infrastructure as complex as the Everglades'.

"I bet you there are less than 10 percent that are not modified," says Garth Redfield, for forty years an environmental scientist, first at the National Science Foundation, then at the South Florida Water Management District. To glimpse a river or stream that remains pristine he suggests a trip into some of the nation's highest and most remote mountains, like the Rockies or Sierras.

The Great Lakes are composed of five of the world's largest freshwater lakes, together containing some 90 percent of North America's fresh surface water supply, and are managed in partnership with Canada. The mighty Mississippi River, among the world's largest rivers, courses between some 3,500 miles of levees and above some 190 bendway weirs, which are like little dams designed to be topped as they avert erosion. Spillways along the Mississippi help prevent flooding, and dredging maintains an upriver navigational channel. The Colorado River is the most important water source for some 40 million in the American West, and like the Everglades, it has been replumbed into an artificial system that has left the river dying of thirst.

What makes the Everglades unique is its flat terrain. Most water management infrastructure is constructed around the fundamental concept that water flows downhill, but the Everglades are so flat the water flows where the South Florida Water Management District and US Army Corps of Engineers direct it to flow. Engineers, meteorologists, and water managers monitor water levels and weather twenty-four hours a day. They refer to computer models and data as they manage hundreds of water control structures throughout the region. The complexity is comparable to that in the Netherlands, where another of the world's most innovative water management infrastructures, in that densely populated country situated mostly below sea level, is admired across the globe.

Meddling with these waterways has made way for modern life as we know it. Among the worst consequences has been what is called stormwater, or the rainwater that falls on our homes and communities and flows away through a labyrinth of gutters, drains, canals, retention ponds, sewer systems, and other water management infrastructure. The labyrinth ends at our waterways, where the water arrives burdened with all of the nutrients and pollution it can carry from our dirty roadways, fertilized crops and lawns, and old, leaky septic systems. In our waterways the nutrients and pollution wreak havoc as they did that summer of 2018 in Florida, the nutrients nourishing toxic algae blooms as they do crops and lawns as part of fertilizers.

A FEW WEEKS BEFORE MY Lake Okeechobee boat ride with Iglesias, the Jacksonville District's Water Management Section convened one of its periodic conference calls with government representatives, scientists, and the public.

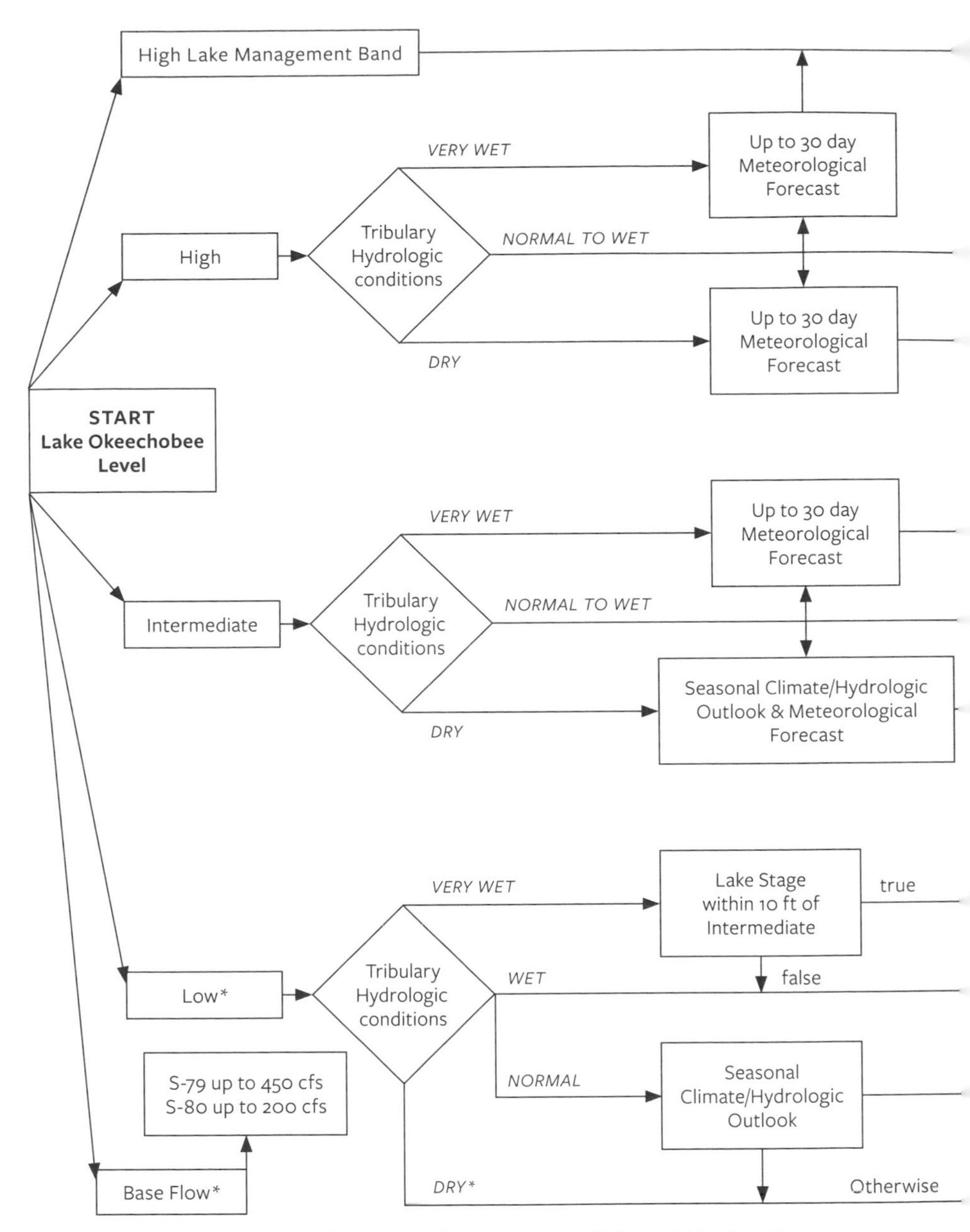

* Very Dry Conditions may require that releases to tide (estuaries) be discontinued.

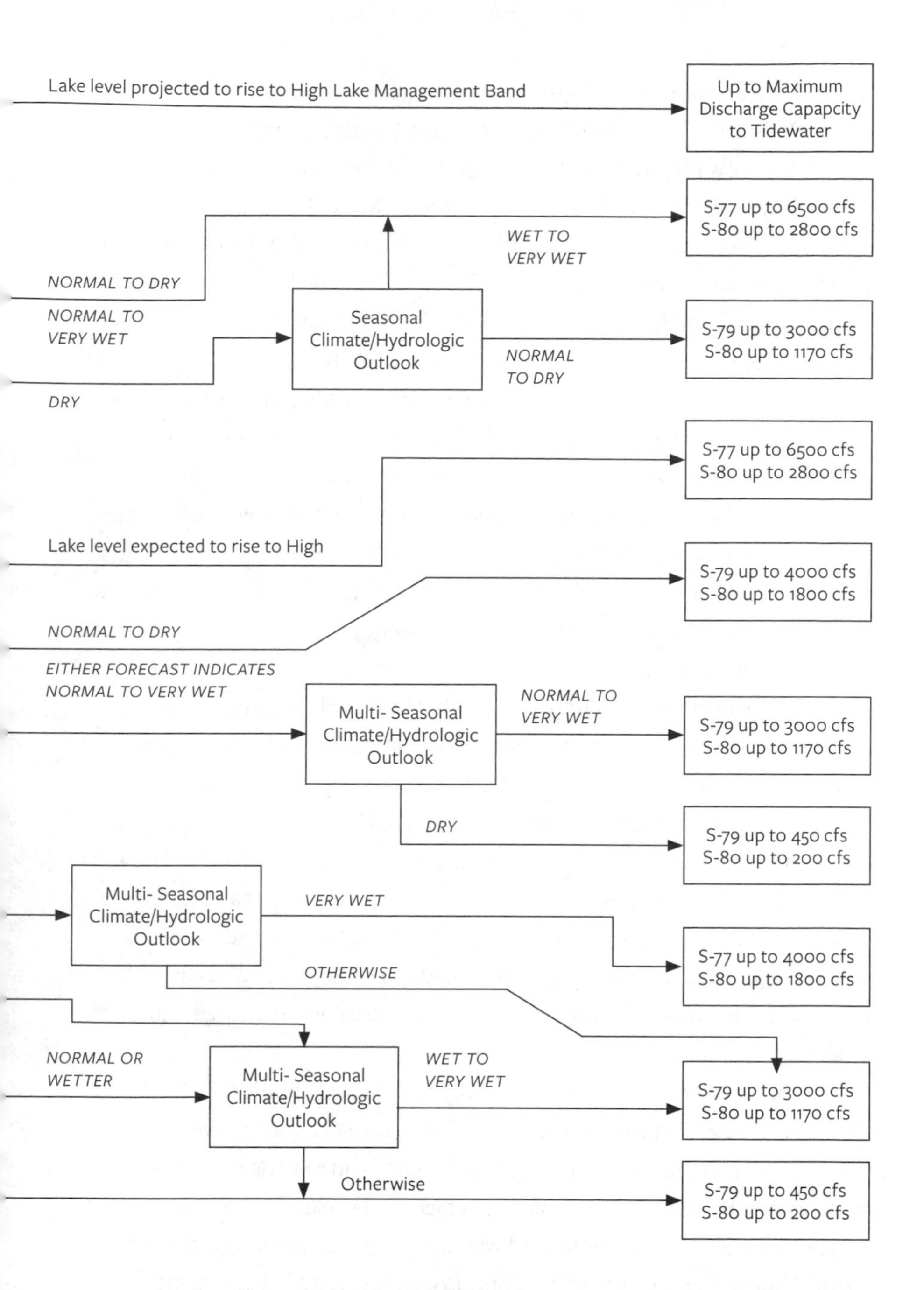

Lake Okeechobee Regulation Schedule (LORS): a decision tree for determining how much water to discharge from the lake (in cubic feet per second, or cfs) based on the lake level, how much water is flowing into the lake, and forecast precipitation

Courtesy of the US Army Corps of Engineers

Dialing in were some two dozen representatives from government agencies, including the Environmental Protection Agency, National Oceanic and Atmospheric Administration, US Fish and Wildlife Service, Florida Department of Agriculture and Consumer Services, Florida Department of Environmental Protection, and Florida Department of Health, as well as local governments along the lakeshore and the St. Lucie and Caloosahatchee Rivers. I called in from the newsroom at WMFE (NPR Orlando), where I worked as an environmental reporter. The lake was continuing to rise only a month into hurricane season. Everyone knew what was coming. It was inevitable.

The civil engineers of the Water Management Section were faced with a critical decision: whether to initiate flows of water from the rain-swollen lake to the St. Lucie and Caloosahatchee Rivers and sensitive estuaries where the rivers spill into the Atlantic Ocean and Gulf of Mexico.

Lake Okeechobee is a moderately eutrophic, shallow freshwater body that has suffered ecological change because of nutrient pollution. Not only do pollutants flow into the lake from Orlando and Central Florida's cattle ranches, but sugar growers used to pump their water into the lake, until problems in the lake prompted a lawsuit that forced the water south into the Everglades. Over the years all the nutrient pollution has created ideal conditions for cynobacterial blooms that have been documented in the lake since 1970. In 2018 the rising water level threatened a lake ecology dependent on shallower conditions and also the lake's aging dike, which held back the Big O from inundating the farming communities to the south. But the releases of water would allow the lake's toxic algae to spread to the St. Lucie and the Caloosahatchee and would also further disrupt the estuaries' delicate balance of fresh- and saltwater.

On the call the local government representatives implored the Water Management Section to hold off on the flows. Already in southwest Florida there was blue-green algae in the Caloosahatchee and red tide in the gulf. The locals worried about the economic impact as hotels fielded cancelations from vacationers turned away by the foul water. Noah Valenstein, secretary of the Florida Department of Environmental Protection, cited the state of emergency declared the day before by Governor Rick Scott, a Republican engaged that election year in a bitter and ultimately successful campaign to unseat

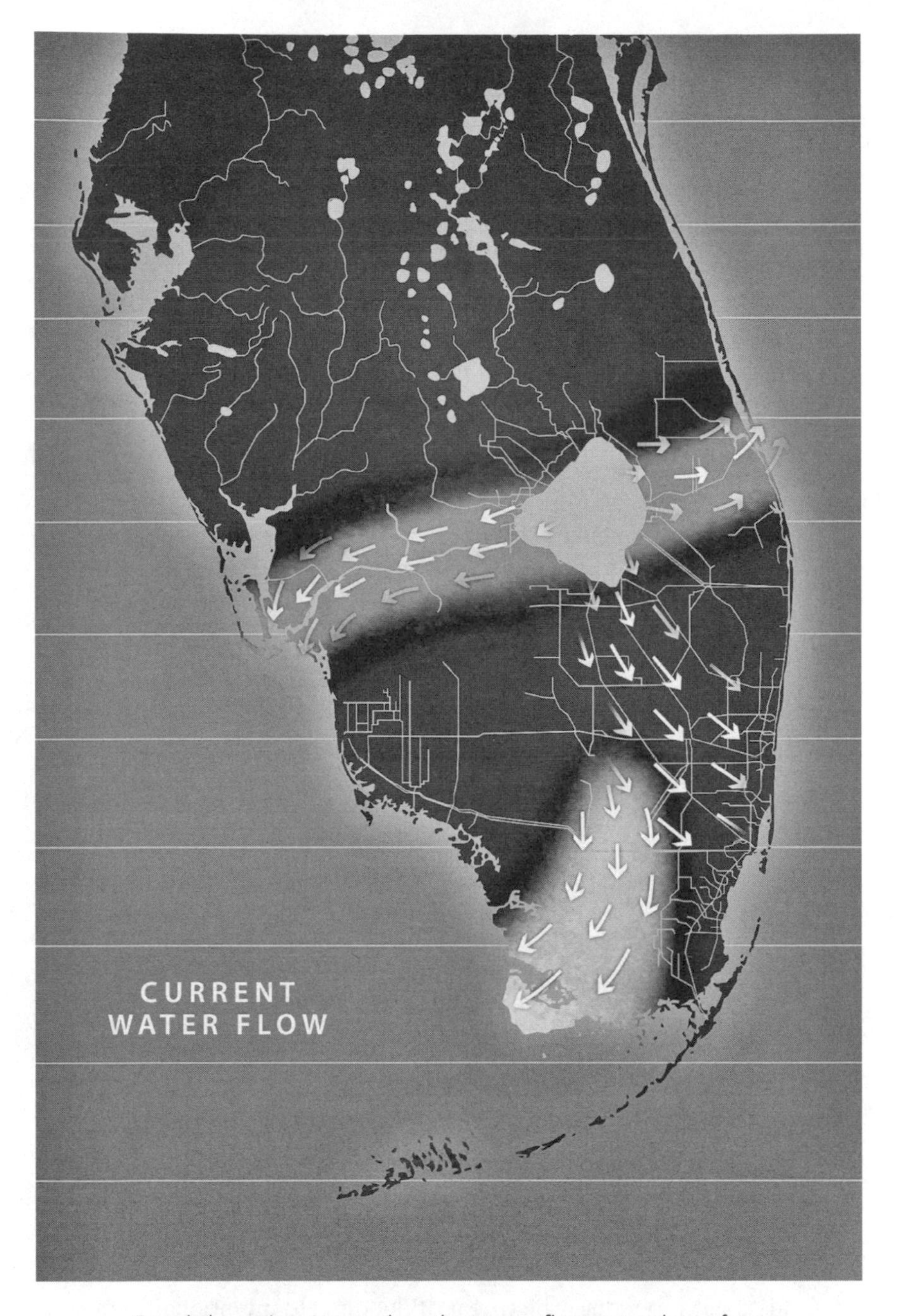

Everglades region. Arrows show the current flow east and west from Lake Okeechobee, rather than south, as the water historically coursed.

Courtesy of the Everglades Foundation

Democratic senator Bill Nelson. A representative of one environmental group delivered a bleak report on the Caloosahatchee and asked whether it might be possible to spare the rivers and estuaries of more toxic algae by moving lake water temporarily onto the farmlands of the Everglades Agricultural Area.

"The entire river system is involved in this algae bloom," she said.

In the end the lake water's rise so early in hurricane season left the civil engineers of the Water Management Section with little choice. Days later the US Army Corps of Engineers' Jacksonville District announced the flows would occur in pulses simulating rain to try to minimize the environmental impact. Even so, the decision was explosive during an election year and two years after a previous outbreak of blue-green algae that was so awful the beaches were slimed over during Fourth of July weekend.

The gates opened, and Lake Okeechobee's dark water spread across the bright blue-green of the St. Lucie's and Caloosahatchee's estuaries like a shadow. Businesses suffered, and sickened Floridians who had come in contact with the toxic algae turned up in emergency rooms. Meanwhile the red tide expanded its grip on the Florida peninsula. Tons and tons of dead fish washed up on beaches, and manatees, sea turtles, and George Barley's beloved tarpon went belly up. The governor declared a second state of emergency in response to the red tide.

"Today, our state is once again facing a crisis from water releases controlled by the US Army Corps of Engineers," he said.

A few weeks later I talked with John Campbell, a Jacksonville District spokesperson, about the idea of holding lake water temporarily on the farmlands of the Everglades Agricultural Area. He said the idea would be considered.

"One of the reasons that the whole system was built was to reduce the risk of flooding on private property," he told me. "I understand why the question was asked, and I understand why many people are feeling economic impacts. But no one is sustaining property damage currently, and so that is a very serious step when you're asking a government agency, who one of their missions is to reduce the threat of flooding, to intentionally consider flooding land."

It did not happen.

MARY BARLEY WAS FRUSTRATED.

In August 2018 I went to see Mary in Islamorada. Nearly a year had passed since Hurricane Irma had slammed into the Florida Keys, yet business signs were still blown out and roof shingles remained torn off. George's beloved Florida Bay similarly had not fully recovered. The Category 4 hurricane had forged a path of destruction through the keys and also the bay, uprooting large swaths of seagrass and leaving behind a barren bottom beneath turbid water. All of these months later, the water quality at these bald spots remained green and muddy, like pea soup.

The bay also continued to thirst for freshwater, even as Floridians farther up the peninsula squabbled over what to do with too much of it.

"It's one of the most frustrating things," Mary told me.

We were seated at her monkeywood dining table, fashioned from a Key West tree downed during a previous hurricane. A Murano glass fruit chandelier dangled from the ceiling above. Behind me on the sky-blue wall hung bonefish, permit, redfish, tarpon, trout, all part of Mary's expansive taxidermy fish collection, which had come with the house. In the adjoining living room above cream-colored leather couches hung paintings that could have been of George and Mary Barley in Florida Bay, casting from a flats boat or reeling in a tarpon. French doors proffered a sunny view of Barley Basin and beyond that, Florida Bay.

Draining the Everglades had left Floridians with not only toxic algae but a bizarre and perpetual mix of too much freshwater and not enough of it. Twenty-six years had elapsed since George's fateful birthday fishing trip. I asked Mary what he would think of it all.

"He would be so angry. He wasn't really an angry guy. I think his head actually would be exploding. I think he would be calling everyone he could think of and flying people around," she said. "I know what he would be saying: 'I told you so. I told you 20 years ago, 30 years ago this was going to happen.'"

5

THE CAMPAIGN BEGINS

Unlike the natural waters in many states, Florida's are held in a public trust and cannot be bought, sold, or owned by any private entity. The state's Department of Environmental Protection and five water management districts are charged with enforcing the laws aimed at keeping these waters clean and sparkling. Except in September 1993 it seemed to George Barley as though Big Sugar owned the Everglades and Florida Bay, and there was little anyone was doing about it.

Not only did the Everglades Agricultural Area, situated south of Lake Okeechobee, staunch the river of grass's historic flow. The world-class water management infrastructure serving as life support for the Everglades drained and irrigated the cane fields on an upside-down schedule that left them dry when they otherwise would have been wet and wet when they would have been dry. Floridians paid for all of this with their taxes, plus the cleanup of the nutrient pollution flowing from the fields into the watershed that was a crucial drinking water source.

"It wasn't just that the sugar industry had so much influence in government that they could manipulate decisions," says Eric Draper, for thirty years a lobbyist for Audubon Florida before he was named director of the Florida State Parks System. "They manipulated government to their benefit and to the detriment of both the environment and ordinary citizens."

George concluded the state's leaders were too corrupted by sugar growers to address the Everglades' and Florida Bay's problems. One evening he found a quiet moment with Mary in the sitting room with a fireplace off their bedroom in Orlando. It was where the couple often held important discussions. George told Mary he had decided to retire early and invest himself full-time in a campaign that would take the problems to voters in a statewide referendum. George prepared Mary for his travel and what he believed would be a tough battle ahead.

"He could get threats, and he didn't know what was going to happen," Mary says George told her. "I don't know that he was actually physically threatened, but all our friends told us to be careful. These guys were tough. Took no prisoners."

GEORGE DECIDED HE WOULD call it penny-a-pound.

He established a political action committee, named it Save Our Everglades, and hired three attorneys to perfect the referendum language for the November 1994 ballot.

The referendum proposed a state constitutional amendment that would raise money for Everglades restoration by imposing a penny fee on every pound of sugar raised in Florida. Growers would pay $35 million a year at the annual production rate at the time of 1.7 million tons, generating $875 million over twenty-five years, more than twice what they would pay under the Interior Department plan. The money would be held in a trust fund managed by a five-member board of trustees appointed by the governor. The trustees had to be Florida residents with experience in environmental protection and could not serve in an elected public office at the same time. George didn't want to waste time. The fee would take effect the day after voters approved it.

To get the referendum on the ballot George needed more than 429,000 signatures from voters in at least twelve of the state's twenty-three congressional districts. The signatures had to be verified by elections officials in each county where they were collected. Once 43,000 signatures were verified in three counties, the state supreme court would rule on whether the language was constitutional.

George sank $10,000 of his own money in the campaign. His good friend Paul Tudor Jones, a Wall Street billionaire, gave $876,000. Other wealthy Central Florida supporters included corporate attorney Thom Rumberger and hotelier Harris Rosen. Audubon Florida, Clean Water Action, and other environmental groups joined the effort. Nat Reed, a South Florida Water Management District board member and former undersecretary of the US Department of the Interior, also got involved. Mary estimates George spent half of his time traveling.

"He was going to the Keys and Tallahassee and Everglades National Park, going to the corps in Jacksonville and the South Florida Water Management

George Barley in the early 1990s in Arthur R. Marshall Loxahatchee National Wildlife Refuge, part of the Everglades ecosystem

Courtesy of Mary Barley

District," she says. "He was working 24-7, probably. He was calling, faxing. In those days we probably went through a ream of paper a day, if not more."

George hired Hank Fishkind, among the state's leading economic consultants, to evaluate the Interior plan. Fishkind found the plan greatly exaggerated sugar growers' contribution and estimated their real share of the cost at 29 percent compared with the public's 71 percent. George believed this was public money that could be spent on Florida Bay. He believed the growers were not paying their fair share and, in personal notes, accused the state of engaging in a "cover-up" aimed at concealing this fact from the public. He amassed volumes of government documents supporting this conclusion. He hired a campaign manager, lobbyist, and communications director. He also hired a company specializing in petition drives to help volunteers collect signatures. Conference calls were held as often as twice a week. A lot of the work was done out of Barley's office.

"I've laid this out kind of like you would do a real estate deal," he told one newspaper. "What you've got to do first is get the language qualified. Usually that takes about two weeks. We got it done in 20 minutes. We just had someone from Fishkind's office walk it into the elections supervisor's office, hand it to them and say, 'I'll wait.'

"Now we have 30 days to get 40,000 signatures. We'll do that. Then it goes to the Attorney General, and he sends it to the Florida Supreme Court for review. At that time the sugar industry will send in about 40 ex–Supreme Court justices they've hired to argue that our initiative is unconstitutional.

"They'll lose that argument."

IN NOVEMBER 1993 George Barley wrote in a memo to a business associate:

> The ballot initiative is off and running and so far doing well. Petitions are being gathered at a rapid pace, 3,000 to 5,000 a day. We will have the qualifying number this month and the Attorney General, who has given us an enthusiastic audience, will present it to the Supreme Court for their review on the single subject matter around the end of the year. Our lawyers feel confident it will not

only pass that review but withstand the other legal challenges ahead. The good news is when it passes the fee will be collected January 1, 1995 and sugar's constitutional challenges will be expedited and over with before the year 1995 is over and the money will become available for Everglades restoration. This offers a substantial impact on the settlement negotiations underway where sugar is behaving in typical fashion now tying up two assistant secretaries of Interior in addition to the Florida Department of Environmental Protection and the South Florida Water Management District. Fanjul is reported over to the White House this week pulling strings.

We have formed two organizations. 1) Save Our Everglades Inc.—a C4. This organization is getting the petitions signed and building a staff for the initiative campaign. It has a substantial budget toward which Paul [Jones] has made a significant commitment. This is the hardest money to raise, non deductible. We are raising other money in a variety of ways. I have contributed as well. Our goal is to raise $2 million from sources other than Paul.

2) The Everglades Trust—a C3. This organization will engage in selling the Everglades. It will commission a number of studies, such as one on the economic value of the Everglades to Florida. It is proposed this study be done by an outfit in Washington, Resources for the Future. We may also do a study on the amount of water and other subsidies for the sugar growers apart from price support and import quota subsidies. Two studies on the latter subject have already been done for this organization and these two studies cost approximately $25,000. Another study for this organization done by economist, Dr. Hank Fishkind, was on an economic analysis of the Statement of Principles . . . [and] cost about $6,000. The Everglades Trust will have additional expenses involved in raising the awareness of citizens to the importance of the Everglades to them. Another possible study will be on the importance of the Everglades to water supplies in south Florida.

Paul and I need your help in seeking funding for both of these organizations. We need to keep all of Paul's money in SOE, Inc the non deductible organization as that is where it is so hard to raise

> money. I have initiated some conversations with the Dunn Foundation but I need a lot of guidance in how to go about this. Please put the best part of your brain power on this and counsel me. We are hopeful NF&WF [National Fish and Wildlife Foundation] together with its family of associated foundations will do much of this for us.
>
> Meanwhile Florida Bay gets worse. 100,000 acres of seagrass gone and an algae bloom at least 1,000 square miles. Half the juvenile lobsters are reported gone and more seagrass die off found on the Atlantic side. Charter boat captains report alarming visibility conditions on the Atlantic side and Key West citizens who went out lobster diving were shocked at the drop in visibility that far down. Sugar is calling us scorched earth environmentalists appealing to mob rule, kooks and liars. We are receiving growing and widespread support, especially from Florida business.

As 1993 drew to a close George traveled to the Everglades Agricultural Area to visit with a group of farm women at the historic Clewiston Inn. He wanted to explain his plan for saving the Everglades with a penny-a-pound fee—or tax, as the growers called it—on sugar. Among the women at the meeting was Vee Platt, a seventy-eight-year-old widow who had lived in Clewiston for most of her life. Her late husband had established a sugar farm, and she still had 650 acres in production.

Platt would tell a newspaper she thought George was nice but lacked common sense about sugar farming and life. She stayed in touch with him, though, and sent him a fax every time she read a newspaper article about his proposal. George would write back; he even penned a note for Platt's eleven-year-old grandnephew describing how sugar farming was harming the river of grass. The correspondence disturbed the boy's parents as the youngster began raising questions about whether farmers would have to abandon the region.

"Mother Nature is always battling you with worms and bugs and bad weather," Platt told the newspaper. "These environmentalists are even more stressful than Mother Nature. And confused! They think God is down here," she said, pointing at the Earth, "rather than up there."

And she pointed toward the heavens.

THE GROWERS' ATTORNEYS filed hundreds of pages of briefs describing what they thought was wrong with the Save Our Everglades proposal. Their primary argument was that the proposal violated Florida laws limiting citizen ballot initiatives to a single subject. The attorneys contended that the fee or tax, along with an independent board with spending powers, represented two separate issues. The proposal's ballot language would confuse voters, they added, and the language unfairly accused growers of polluting the Everglades.

"For the record, 'the sugar cane industry' emphatically denies that it has 'polluted the Everglades,'" stated one brief filed by the Sugar Growers Cooperative of Florida.

Almost single-handedly, George Barley had upended what at the time was Florida's biggest environmental debate ever. His penny-a-pound campaign threatened an imminent settlement of five years of litigation over how to save the Everglades and who should pay for it.

The Republican land developer also had united the state's splintered environmental groups with wealthy businesspeople who had the means to advance their agenda.

"Typically in the environmental community you have the problem of a lot of idealistic people without the political, economic, and scientific skills to bring together an effective campaign, and I think because [George] was both a developer and used to dealing with engineers and used to dealing with political situations, he had an acumen to add that value to those of us that were laboring on the environmental side who didn't necessarily have those skill sets at that time," says Charles Lee, an Audubon Florida advocate of nearly five decades.

"Being a developer and being faced with requirements to comply with environmental regulations as a developer, he couldn't understand why when, if he built an office or apartment complex in the Orlando area he had to set aside maybe as much as 25 percent of the land to build storm water retention ponds. He didn't understand why the sugar industry south of Lake Okeechobee could build and maintain sugar cane fields without having what would effectively be the same thing. . . .

"George was arguing, 'OK, the government is providing them with this big subsidy. Government is running their pumps for them. Government is not requiring them to build their own retention ponds, and therefore we need to do something to collect some money from these people so that they'll be paying their fair share.'"

George also helped seal into Florida environmental history a lasting narrative in which sugar growers are the villains to the Everglades. Few of the state's ecosystems are victims of something so easily defined. Growers characterized George's penny-a-pound fee as an impossible financial burden that would put tens of thousands out of work, and they quietly began pushing antitax amendments. Some went as far as to question the morality of the fee.

"Do we really want to live in a state where people can get together by popular vote and say who is guilty or not?" asked Bob Buker, senior vice president of U.S. Sugar Corp., which contributed $1 million toward the antitax amendments. "It means the majority can beat up on an unpopular minority. We just happen to be a politically correct minority to beat up on."

IN FEBRUARY 1994 Governor Lawton Chiles introduced an Everglades cleanup bill named the Marjory Stoneman Douglas Act, after the author-turned-activist whose book *The Everglades: River of Grass* had introduced the elusive watershed to readers nationwide. The growers hired three dozen lobbyists to work the measure, including the governor's former chief of staff and two former house Speakers. George approached Florida's best-connected law firms about representing the Everglades Coalition, but most already represented the growers, and when he found one that didn't, a grower landed the firm a few days later. The growers also organized support from other agriculture groups, business interests, and labor unions, which warned lawmakers that strict phosphorus standards would endanger mill jobs.

The legislation delayed final phosphorus standards until 2006. Growers were responsible for less than a third of the $700 million the measure included for the now 40,000 acres of filter marshes that were expected to get the phosphorus level down to 50 parts per billion. The measure included no funding for getting the level down to 10 parts per billion. US Attorney Dexter Lehtinen said the bill would gut the water quality agreement he had negotiated in court. Douglas, then age 103, wrote to Chiles herself to demand that her name be removed from the measure. "I disapprove of it whole-heartedly," she said. The bill was renamed the Everglades Forever Act.

That spring the legislature approved the measure, billed at the time as the largest environmental restoration project in history. On a boardwalk of Everglades National Park, Chiles signed the bill into law. The legislation ended thirty-six lawsuits over the Everglades and Big Sugar. Environmental groups resoundingly denounced the measure, arguing it put the burden on taxpayers. The law set up a tax of 10 cents per $1,000 for taxable property in the South Florida Water Management District, which spans sixteen counties in Central and South Florida. The residents would be taxed for ten years, and the tax was expected to raise some $12.3 million annually. Sugar growers said they were not happy with the bill, but they did not plan to protest it.

"We hope we can survive," said Bob Buker of U.S. Sugar. "I hope this resolves the issues. It's a very rigorous bill."

George warned that Big Sugar supporters would pay a political price.

"I'm determined to make this not only a national issue but an international one, and I have the money, resources, connections and determination to do it," he wrote to Joe Browder. "I do not want to reveal all of our strategies in this campaign, but there are unpleasant political surprises in store for those on the wrong side of this issue."

Days before the ceremony George wrote about his frustration in an op-ed he had published in the *Palm Beach Post*:

> There is only one good provision in the much-touted Everglades bill on its way to Gov. Chiles. That is an amendment put in at the last minute by Sen. Curt Kiser, R-Dunedin, and Rep. Sandy Safley, R-Clearwater. Their amendment ensures that the South Florida Water Management District will be able to use eminent domain to buy a portion of the Frog Pond in Dade County—part of the land that is desperately needed to start returning clean water to a dying Florida Bay.
>
> It was a good amendment, but lobbyists for the governor's Department of Environmental Protection and Big Sugar's friends in the Legislature fought hard to keep it out of the bill.

Legislators could have passed a bill that finally ended the litigation, made the polluter pay and expedited the cleanup. Instead, they made even more concessions to the sugar industry. Why? Because lawmakers are too indebted to sugar's lucrative campaign contributions to take on the industry. It's a lot easier to stick it to taxpayers and cover it up with pronouncements that "we can't wait any longer. Restoration is finally starting." Funny, how we hear these refrains most often during election years.

In his State of the State speech on the opening day of the session, Gov. Chiles did not mention the issue, but the sugar barons were already at work on an Everglades bill. Why wouldn't they when they were again on the courthouse steps of a case they would surely lose? They pulled the same trick last year with the Babbitt agreement, which they walked away from after delaying long enough to put off paying for another year. To deal with the Legislature, sugar lined up almost 40 lobbyists to push the bill through on their terms. Near the end of the session, sugar lobbyists crowded the governor's office until after midnight to work on the bill in a secret March 30 meeting. Environmental and public-interest lobbyists were excluded from the meeting. At 1:59 a.m., the final bill was printed and readied for an 8 a.m. committee meeting. So much for government in the sunshine.

The bill awaiting Gov. Chiles signature provides the following:

- A delay until 2003—nearly a decade—before the state finally decides how much pollution the sugar companies can discharge.

- A delay until 2006 in enforcement of water-quality standards for some polluting growers. These delays throw out the Everglades protection that could have been achieved in a 1991 federal court settlement, which required sugar to reduce pollution by 1997 and clean water by 2002.

- Florida taxpayers paying most of the cost of a delayed cleanup of the sugar-company pollution. The agribusiness corporations that dump their pollution into the Everglades would pay only $12 million a year toward the $950 million–plus cost of a delayed cleanup while the public would have to pay more than $23 million a year, as well as $33 million in Preservation 2000 money—tax dollars supposedly set aside to purchase park and recreation lands. On top of that, the project also uses federal taxes to clean up sugar's pollution. The bill shifts to the public more than $400 million in costs that should be borne by sugar. Sugar's share comes to just $233 million. And if even more money is needed, the additional costs would come from the public because sugar's liability is capped.

- Opportunities for the sugar industry to go back to court to postpone final cleanup of their pollution beyond the generous deadlines included in the bill. These extra delays were built into the bill by sugar's lawyers even though the main justification of U.S. Interior Secretary Bruce Babbitt and Gov. Chiles for spending our tax dollars on the cleanup was their fear the sugar companies would sue to stall the cleanup if they were required to pay their full share of the costs. This is one of the most serious flaws in the bill. Why didn't the Legislature settle this once and for all? The obvious answer: Sugar's lobbyists got to them.

State and federal officials and now the Legislature have surrendered to the big sugar corporations that are polluting the Everglades. The terms of surrender are simple: Allow sugar to continue breaking the law, polluting the Everglades, wasting water desperately needed by both the Everglades and South Florida's urban economy and making the public pay most of the costs.

This is neither surprising nor unpredictable. Sugar's influence is so pervasive that even proponents of good government like Gov. Chiles and Sen. Bob Graham, D-Fla., are all too easily caught in the industry's web. In July 1990, when environmental groups were working with Sen. Bill Bradley, D-N.J., to reduce federal price supports for sugar, Sen. Graham opposed a proposal to link price supports to a guarantee of water quality. Responding to a recommendation by Everglades Coalition groups that the subsidy be curtailed, Sen. Graham said:

"Before Congress considers such a legislative proposal, I believe it is only fair to give the Florida sugar growers an opportunity to fulfill their commitment (to clean water). Thus, I would like to see if the outlines of a comprehensive water-improvement management plan cannot be hammered out among all of the interested parties in the next few months." . . .

Sen. Graham's "few months" have turned into nearly four years under the sugar industry's well-practiced strategy of delay. The "Everglades Forever Act" ensures only that we will be debating the issue forever—and that's the industry's game plan.

Fortunately, Florida voters don't have to wait forever for the Everglades to be delivered from the political morass that ensnares its future. More than 420,000 citizens have signed our petition for a constitutional amendment that lets the voters do what their political leaders have not had the courage, candor and guts to do—look the sugar barons in the eyes and tell them they have to change their ways.

Passage of the Save Our Everglades amendment will do two things that will forever change the way Florida cares for its most important and endangered ecosystem. It will tell the politicians that "business as usual" through bad back-room deals, such as this year's Everglades bill, won't be tolerated anymore. And it will finally make the polluters—instead of innocent taxpayers—pay their fair share of the cost of a real and prompt cleanup of the Everglades.

WEEKS LATER, IN MAY 1994, Jon Mills was in his office when the phone rang.

"You've got to be kidding," he said. He hung up. Then he called Mary Barley.

Mills, a University of Florida law professor and former state house Speaker, had written the penny-a-pound referendum language, and now the state Supreme Court had declared it unconstitutional. In a unanimous opinion the court said the language smacked of "political rhetoric" and addressed more than one subject. They protested that not only would sugar growers be taxed but a "virtual fourth branch of government" would be created in the five-member board of trustees, which would take on both an executive duty in addressing the pollution and a legislative duty in handling the tax money. The judges also believed holding growers liable was a judicial duty.

I asked Mary whether she remembered talking to George that day.

"No, but I'm sure he cussed," she said.

"All of the signatures we had collected on that were not worth anything, and all of the money we had spent was for naught," Mary said. "Now we were back to zero."

George Barley put out a news release that said the campaign had collected some 600,000 signatures from supporters:

> They signed it because the sugar industry has given so much money and favors to their friends in government that our own elected officials put sugar's concerns over what is fair and just to the taxpayers of this state. The bill Governor Chiles approved and the legislature passed this year committed the taxpayer to pay $400 million to clean sugar's pollution . . . and in this country people expect polluters to pay to clean their own pollution.
>
> This is only one battle in a war to save the Everglades. The war is not over, and the army of people who joined Marjory Stoneman Douglas in signing our petition is not going to stop until the war is won. In adversity lies opportunity, and the opportunity before us is to help the people's movement grow and swell until government bends to the will of the people. We will not stop until that happens.

These despoilers of the Everglades and their political friends would do well to listen to the winds of change in Florida. Those winds say, restore the Everglades. Make the polluters pay, not the people.

6

THE BARLEYS

Chris Ball arrived at the job interview in January 1995 wearing a dress and heels. She knocked on the front door of the Barleys' home in Orlando and, her eyes wandering, noticed a robin in the grass of the front yard. It was odd, she thought, to see a robin like that in the dead of winter, and as George opened the door and welcomed her inside she hoped the bird would bring her luck.

George Barley was looking for an assistant to help with his Everglades campaign, and his friend Thom Rumberger, a well-known and influential Central Florida attorney, environmentalist, and political strategist, had recommended Ball, one of the kindest people I ever have met. George showed Ball a chair and then, before saying much else, presented her with a slide show on the river of grass, complete with before and after photographs and aerial views of the degradation.

George had given the presentation many times. He took it with him whenever he traveled to the state capitol in Tallahassee. He shared copies with the environmental groups so they could use it in their own advocacy. Ball was impressed.

"He was animated and angry because I think every time he saw that presentation himself it just recharged his anger and made him motivated. I don't know who he was more motivated at the most, Big Sugar or our government for allowing it to go on," she says. "He was explaining to me, 'This is what I'm looking for, and this is what we'll be working on.' So that was the interview, and when he was done it was like, 'What do you think? Do you want to be on board?'"

George didn't want to waste time. He told Ball she could start the very next day.

"I wanted to do whatever part I could play in it," she says. "He got me all excited."

GEORGE BARLEY WAS BORN May 29, 1934, during the Great Depression in Mayport outside Jacksonville, a seventh-generation Floridian with family roots tracing to Napoleon Bonaparte Broward. Elected as Florida's governor in 1907, Broward helped initiate the draining of the Everglades after having run on the campaign promise that he would "reclaim" the swamp for farmland.

On his mother's side George Barley was from a family of boat builders and commercial fishermen. His father was from Manassas, Virginia, and sold fertilizer, and after the family moved to Winter Garden outside Orlando his father served as mayor and a city commissioner.

George had two brothers and a sister, and their childhood was characterized by financial hardship and food lines. He was a skinny kid with big ears whose growth was stunted by illness, although as an adult he would grow to be 6-foot-2. He loved reading, especially Ernest Hemingway, and loved the outdoors. When he was twelve his father bought him a .22-caliber rifle, and his uncle taught him fly-fishing.

"I went to the woods, and in a sense I never came back," he said.

George and his brother were the first in the family on either side to go to college. George went to Harvard University on a scholarship. He graduated in 1956 with a degree in English literature and went straight to work at the American Can Co. in New York. He returned to Florida two and a half years later with a wife, two girls, and, he said, "$50 in my pocket." He settled in

Titusville and eventually got into real estate right as Florida's Space Coast began to flourish, after President John F. Kennedy announced in 1961 that Americans would rocket to the moon. He later moved to Orlando just as Walt Disney revealed in 1965 his plans for Disney World, and George's companies brokered many of the major real estate transactions that would transform Central Florida from backwater to cosmopolitan. Along the way his family welcomed a third daughter.

George Barley was a strict parent. He expected his daughters to earn straight As at school and do chores at home, and he loathed laziness. His wife, Shirley, a homemaker and devout Christian who later would travel to the Middle East as a missionary, shared his energy, and together they could be explosive. In the evenings the family hosted big dinners with extended family, or he invited over other developers and bankers to drink and play cards. During summers the family traveled the country in a Dodge motor home, fly-fishing and hiking along the way.

"My mom was like June Cleaver, and my dad was like James Bond," says Lauren Barley, the family's oldest daughter. "To have him as your dad was just like the most amazing experience ever. He would open your eyes to the world, no boundaries."

THE GAVEL SOUNDED, finalizing the vote with its percussive strike. George Barley lifted the barrel of his finger, took aim, and released the trigger of his thumb, firing at state senator Robert Wexler.

The state senate's Finance and Tax Committee had voted 7–6 to approve an Everglades measure opposed by environmentalists. George was expressing his dissatisfaction with Wexler's decision at the last minute to change his vote in favor of the measure. George was also implying what the ramifications might be for Wexler's reelection. George "was the first one that I remember really starting to use political contributions to kind of reward friends and punish enemies," says Eric Draper of Audubon Florida and the Florida Park Service.

George was smart and a visionary, and as his business grew, so did the breadth of his professional and political connections. He hunted with investors and political leaders like the mayor of Orlando. He traveled Europe for months at a time. He served as a board member or commission member for some two dozen business, conservation, and government organizations.

"He knew, like, the who's who on the planet," says daughter Lauren.

He helped fund raise for Senator Bob Graham, a Democrat, and after Graham was elected governor he appointed George chair in 1983 of the state's first Marine Fisheries Commission, which later would be combined with two other state agencies to form the Florida Fish and Wildlife Conservation Commission. At the Florida Marine Fisheries Commission, George pushed for a congressional investigation of delays to implementing emergency catch quotas for the then dwindling king mackerel. He charged that the National Oceanic and Atmospheric Administration (NOAA) and the National Marine Fisheries Service had shown "serious mismanagement" in caving to political pressure on the issue from other gulf states, namely Louisiana and Texas. NOAA countered that internal disputes among the states' advisory councils rather than federal foot-dragging had caused the delays. George wrote to numerous US senators and representatives and sought support from groups such as the National Audubon Society.

The experience primed George for his appointment in 1992 to the newly created Florida Keys National Marine Sanctuary Advisory Council. NOAA made the nominations and checked them with the White House. Council members also underwent federal background checks. That July he and Billy Causey, the sanctuary's manager, were hung in effigy by fishermen and treasure hunters angered by a NOAA ban on treasure hunting in the sanctuary. The protesters, part of a group called the Conch Coalition, were also alarmed that the National Marine Fisheries Service was considering two twenty-mile-wide no-fishing zones off the Keys, which would have been coordinated with the sanctuary, and the protesters strung up the likeness of a third, unidentified sanctuary official, too.

MARY FYE WAS BORN June 1, 1946, in Oconto, Wisconsin, a small town on the west bank of Green Bay. She was the middle child in a family of five siblings raised by a single mother on welfare. Her father left when she was an infant, and she never knew him. The family lived for most of her childhood with her grandmother in an old boardinghouse with a chicken coup and garden outside, sharing meals around a big table with other boarders. No father figure was ever in the picture. The children walked to Catholic school because the family had no car. They did chores such as cleaning, doing laundry, and picking beans. For fun they went outside and built forts, fished in the river, and played hide-and-seek, although she describes herself as a loner.

"I came from a very poor family," she says. "I never even had a new dress until I was sixteen."

She married at twenty-one and, as Mary Enders, moved to Orlando with her husband, who was in real estate.

"We were not meant for each other," she says. "I can remember walking down the aisle thinking, 'What am I doing?' "

She got a job at a real estate brokerage firm, as an assistant for George Barley's secretary.

EVENTUALLY GEORGE ABRUPTLY LEFT his wife, ending a nearly twenty-year marriage and devastating three daughters, the youngest only seven. The couple divorced in 1972. Mary Barley got divorced at about the same time, and by then was a receptionist at the firm where George worked. A relationship began to flourish that was tumultuous. During the work week they planned weekends together fishing, hunting, or traveling, but they broke up often.

"I never had to have a man to feel fulfilled, probably because of my father," Mary says. "That was probably one reason why he loved me, was my independence."

During the early years of their relationship the Orlando real estate market suffered a recession, and George lost everything. Together the couple rebuilt the business, and Mary sometimes gave money to George to help him out.

"We were great friends as well as business associates," Mary wrote in 1996. "We spent an enormous amount of time with each other, and even after all the years together we still worked and played almost exclusively with each other. We both loved to fish; he hunted with guns, I with my camera. If he was traveling without me he called at least once a day but usually it was more

George and Mary Barley on the River Tweed in the United Kingdom, 1990

Courtesy of Mary Barley

like five, six, even ten times we would talk. We were friends, lovers, business partners, and soul mates. We enjoyed each other, and all our friends knew it . . . and understood it . . . and wished they had it."

The Barleys traveled the globe together. Africa. Cambodia. Hong Kong. South America. They spent a lot of time in Portugal, Scotland, Spain, and especially London, where they had many friends. Then George took Mary to Boston and proposed at a fancy restaurant, but she turned him down.

"I was like, 'Why would you want to do that?'" Mary says. "'Everything is going well. Why would you want to get married?'"

For once George took no for an answer.

He told Mary it was the last time he would ask and that if she ever changed her mind she would have to ask him. The following year she did, and in 1990 they wed, some twenty years after first meeting. George planned the wedding. It was held at the Queen's Chapel of the Savoy in London, and for their honeymoon he and Mary fished on the River Tweed in Scotland.

"George needed space to grow and be the visionary the good lord had given him the gift to be," Mary wrote in 1996. "George was a most unique being and one worth giving full rein to his ability to solve complex public policy. He never let me down nor did he ever let down the people of Florida he represented at times, nor did he ever allow a natural resource to be exploited without trying to right a wrong."

7

THE FANJULS

Monica Lewinsky knew something was wrong. She could hear it in President Bill Clinton's voice.

The White House intern was in her Watergate apartment on the phone with the president in February 1996. She wanted to come see him, but he did not know how long he would be there. So Lewinsky went to the White House anyway and, after presenting a folder she said was for the president, was shown into the Oval Office by a plainclothes Secret Service agent.

In the Oval Office Clinton told Lewinsky he no longer felt right about his extramarital relationship with her and wanted to end it. Lewinsky could continue coming to see him only as a friend. He hugged her but would not kiss her. At one point during the breakup, which would only be temporary, the president got a call from a Florida sugar grower whose name Lewinsky would remember as something like "Fanuli." She would tell special prosecutor Kenneth Starr she thought the president might have taken or returned the call as she left.

The call was from Alfonso Fanjul, who with his brother José, was the largest landowner at the time in the Everglades Agricultural Area. The Fanjul brothers, nicknamed Alfy and Pepe, were Florida's sugar barons, responsible for more than a third of the state's crop. What Alfy discussed with the president during their more than twenty-minute call is not disclosed in Starr's report to Congress, but it likely was about a proposed penny-a-pound fee on Florida sugar for the Everglades.

THE FANJUL (PRONOUNCED FAHN-HOOL) BROTHERS' rise to prominence traces to the nineteenth century when a young Spanish immigrant named Andrés Gomez-Mena arrived in Cuba and found work on the docks of Havana. Andrés soon felt drawn to farming sugar and eventually with his son José bought four mills. After Andrés's death José carried on the family business, serving during the 1930s as Cuba's secretary of agriculture. During this time José's eighteen-year-old daughter, Lillian, married twenty-seven-year-old Alfonso Fanjul, whose father was director of a large sugar trading house. By the 1950s the Fanjul–Gomez-Mena sugar empire dominated the industry in Cuba.

Together Alfonso and Lillian had five children, and the family led charmed lives. Alfy was the oldest and among the richest members of his graduating class at the Catholic all-boys LaSalle School in Havana's wealthy Vedado section. A classmate described him as funny and outgoing and a good student.

Fidel Castro's Cuban Revolution would change the family's fortune. On New Year's Eve in 1958 Alfy and Pepe were watching fireworks at the Havana Yacht Club as word spread that President Fulgencio Batista had fled. Armed militia soon arrived at the Fanjul home and arrested Alfonso. The rebels sealed off parts of the home and commandeered the family's cars. Alfonso was interrogated for hours at police headquarters until finally a woman appeared. "Release him," she said. "He has nothing to do with the Batista government."

At the family headquarters on Avenida de Gomez-Mena in Havana the rebels threw down their machine guns on a conference table. The walls were hung with maps showing the expansive Gomez-Mena and Fanjul holdings in Cuba, some 150,000 acres and ten mills. The rebels circled one of the maps. This will be ours, the rebels said.

Alfonso fled to New York, where he owned some apartment buildings and where a Fanjul sugar trading house was located. He planned to wait out the revolution there and appointed Alfy, then twenty-one and fresh out of Fordham University in New York, in charge, but Alfy was repeatedly shot at and threatened with death. Alfy would lie on his office floor to avoid the gunfire by day and by night would stay with friends, moving from house to house.

Eventually he and Pepe joined their father in New York, just as the United States put an embargo on Cuban sugar and created huge incentives for domestic production. In 1960 Alfonso moved his family to Palm Beach to farm sugar in the newly drained Everglades.

IN PALM BEACH Alfy looked around for some sugar mills to buy. He found two dilapidated mills in Louisiana, bought them with help from partners for some $165,000, and had them dismantled and shipped by barge to Osceola Farms, situated on 4,000 acres in the Everglades near Pahokee. Perhaps as a patriotic gesture the family employed nonunion laborers, especially Cubans who had crossed the Florida Straits by raft and were willing to work for rice, angering neighboring farmers who thought the unpaid labor gave the Fanjuls an unfair competitive advantage.

In 1961 the Fanjuls produced five tons of sugar. Pepe joined the business after earning an MBA from New York University, and the youngest brothers, Alexander and Andrews, soon followed. By the time of patriarch Alfonso's death in 1980 the Fanjul sugar empire was rebuilt. Osceola Farms was producing 50 tons of raw sugar annually and grossing $30 million in sales. The family's holdings also included New Hope Sugar Company and Flo-Sun Land Corp. Alfonso was a member of various exclusive clubs in Florida, New York, and Southampton and was a confidant to dukes, kings, and presidents. His regal white mansion on Wells Road in Palm Beach was a social center.

Then, in 1984, the brothers announced they would buy the sugar holdings of the troubled US conglomerate Gulf & Western for between $200 million and $240 million. The deal included some 90,000 acres of Florida cane and a second mill. It also included a 240,000-acre agricultural empire in the Dominican Republic and the posh Casa de Campo resort, where Michael Jackson and Lisa Marie Presley reportedly were married.

Of forty-one sugar-producing countries the Dominican Republic had, for more than a decade, been awarded the largest foreign quota of sugar exported to the United States. Flo-Sun pushed ahead of U.S. Sugar Corp. as Florida's largest sugar producer. Alfy's net worth in 1985 was estimated at some $400 million, making him Palm Beach's wealthiest resident, ahead of names like Ford and Pulitzer. After living in constant fear of assassination in Cuba, he had become perhaps the most obvious flashy guy in Palm Beach. The family collected opulent mansions, customized Porsches and yachts, and white-gloved butlers. Pepe's red Ferrari Testarossa was his trademark.

Nonetheless accusations of atrocious labor practices involving mostly African Americans dogged the Fanjuls and other Florida growers. After U.S. Sugar was indicted in 1943 for conspiracy to commit slavery (the charge was thrown out because of an illegally drawn jury) the growers began relying on Caribbean cane cutters, and in 1986 some 350 cutters staged a one-day walkout at the Fanjuls' Okeelanta Farm in protest of low wages. The brothers engaged guard dogs to round up the cutters, flew the cutters back to Jamaica, blacklisted them, and replaced them with other Jamaicans. A $51 million civil court judgment in 1992 for cutters' back wages prompted the Fanjuls and other growers to consider alternatives, and a year later the brothers announced a technological breakthrough that would move the industry beyond the labor disputes and mechanize the harvest.

By the time of Alfy's ill-timed Oval Office call the brothers owned 8 percent of the Dominican Republic's arable land, ran the world's largest sugar mill, and produced half the Caribbean country's cane crop. In the United States the brothers bristled at their label as sugar barons.

About "the 'sugar barons' myth," the Fanjuls wrote in a letter to the *New York Times*, published in 1997: "Barons are granted land by a king; we lost ours at the hands of a dictator. In 1959 we fled Fidel Castro's Cuba for the United States, where we started from scratch. It was a modest start with profits guaranteed by no one, and risk, as with all farming, was abundant. Today we're proud to employ more than 3,000 Floridians making an average of $31,000 a year."

ALFONSO AND HIS SONS had started out by giving $200 here, $500 there. In Cuba Alfonso had paid *mordidas*, or bribes, to Batista but had turned down an offer to serve as an ambassador. By 1996 the brothers and their associates ranked among the most prolific political-money machines in the United States, contributing some $3 million over the years to federal candidates in forty-six states, Democrats and Republicans.

More than $360,000 went into the 1992 elections alone, as growers battled the federal lawsuit against the state over the growers' pollution of the Everglades. Alfy cochaired Clinton's Florida campaign and hosted one of his inauguration parties, and his generosity earned him a place at one of dozens of White House "coffees" for major donors. When the Clinton administration announced an agreement the following year that would limit growers' financial responsibility in Everglades cleanup, Alfy was standing right by Interior Secretary Bruce Babbitt's side.

While Alfy was the Democrat, Pepe was the Republican and as deeply involved in the 1996 elections. Pepe served as national vice-chair of finance for Bob Dole's presidential campaign and hosted the Republican nominee for a $1,000-a-person fund raiser.

Perhaps Alfy felt betrayed that February when he learned of a new Clinton administration plan to save the Everglades that included a penny-a-pound fee on Florida sugar. Alfy's call to the Oval Office came the same day Vice President Al Gore announced the $1.5 billion plan at Everglades National Park. Some two thousand sugar supporters had protested the day before at a Miami resort where Democratic officials had met. Alfy had learned of the plan three days in advance, before both of Florida's senators, Democrat Bob Graham and Republican Connie Mack, knew of it. One administration official had told a Mack staffer the plan was not fully formulated.

"I will fight this," Alfy had remarked, "to my last breath and my last dollar."

8

BIG SPECIAL INTERESTS

Florida's debate over Big Sugar and the Everglades spilled into living rooms nationwide in May 1995 when CBS's *60 Minutes* picked up the story.

Seated before an image of sugar pouring from a spoon into a china bowl stuffed with $20 bills, *60 Minutes*' Steve Croft explained that Congress was about to take up a new farm bill and the provision gaining the most attention was the sugar program. Worldwide sugar sells for some 14 cents a pound, he said, but not in the United States, where the government controls the price in the same way the OPEC cartel once controlled the price of oil. Instead Americans buy sugar for some 22 cents a pound as part of a program aimed at ensuring a safe and stable supply while protecting farmers from foreign competition.

"It also guarantees," Croft said, "that Americans will pay a billion and a half dollars a year in higher sugar prices, and guarantees the sugar industry a very sweet deal."

Then images flashed of mechanical cane harvesters in the Everglades Agricultural Area and a twenty-six-room Fanjul mansion in Palm Beach, bright white with a Spanish red-tile roof.

"Don't look for the family farmhouse out here in the cane fields," Croft said.

Few Floridians had seen a cane field, and yet here they were on TV screens across America. Then viewers were presented with an image of George Barley, his brown hair swept back and blue shirt collar undone as he ran a flats boat on Florida Bay. Later in the segment, seated with Croft with the bay as a backdrop, George looked confident, relaxed, but concerned, his furrowed brow adding to the creases of his forehead. He spoke with a lingering drawl from his boyhood in what then was rural Jacksonville, back before the Everglades were drained.

"You see, Steve, money's not an object to these guys," George said. "They get a river of money from the sugar program, from the subsidy program, and that comes on the backs of Americans all across this country."

BEHIND EVERY CANDY BAR and can of soda—in fact, behind nearly everything you consume—is a complex government program of import tariffs and farmer loans inflating the US price of sugar.

This multifaceted program costs consumers more than $1 billion annually through increased food prices, according to various government and independent studies. Back in 1995 George estimated the program cost Florida consumers some $72 million a year, but even today government leaders continue to authorize the program despite the cost because growers pay them to.

In the fifteen years leading up to George's *60 Minutes* appearance, for instance, sugar growers' various political action committees gave more than $10 million in campaign contributions. The Fanjuls and U.S. Sugar Corp. executives spent nearly $600,000, and growers furthered their influence by joining forces with other agricultural interests. Cane was then grown in relatively few congressional districts, in Florida, Hawaii, Louisiana, and Texas (sugar beets were raised in Michigan, Minnesota, Ohio, and a few other places), but when a midwestern senator wanted help from Florida's leadership with a wheat issue, Florida's leadership got help with sugar in return.

"They were capable and still are capable of hiring more lobbyists to work the Florida senate, for instance, than there are senators," says Fowler West, a lobbyist who worked for George. "If they need to they can hire 100 lobbyists for the senate up there [in Tallahassee]."

The situation is not too different today. It means you, the taxpayer, help pay growers to raise cane in the Everglades even as you, the taxpayer, help pay for the environmental cleanup.

It also means the growers' influence extends beyond political leaders in Tallahassee and Washington into some of the most personal aspects of your life, from the food you eat to the water you drink to the environment where you live. The paradigm has been in place in one form or another since shortly after the nation's founding, but in 1995 it looked as though that might change.

The sugar program was up for renewal as part of the five-year farm bill, a massive measure determining nothing short of the foods we eat and how we raise them, and the free market Republicans controlling Congress at the time appeared poised to abandon the program.

Powerful Republicans like House Agriculture Committee chair Pat Roberts and presidential hopeful Richard Lugar were talking of dismantling the system, and President Bill Clinton's economic advisers described the program as "cartel-like." In Florida, Representative Mark Foley, whose district included the Everglades Agricultural Area, said "the status quo is not going to happen," and Representative Dan Miller filed legislation to end the program. Meanwhile a motley coalition of consumer and environmental groups and non–sugar sweetener users like M&M/Mars and Hershey Foods cheered the rising opposition, doling out lollipops on Capitol Hill wrapped in paper printed with the words "Lick Big Sugar."

The opponents pointed to the 8 cent discrepancy between domestic and world sugar prices and also to a 1993 General Accounting Office report that concluded the program disproportionately benefited a small number of growers, especially South Florida's wealthy agribusinesses responsible for the Everglades' plight. The report estimated 42 percent of the benefit went to 1 percent of farms. Republican House majority leader Dick Armey went further, releasing figures showing the Fanjuls had gotten $64 million during the 1993–94 growing season, and U.S. Sugar $55 million. Congressional members like Miller—like Armey, an economist—expressed disgust.

"This is the sugar daddy of all corporate welfare programs," Miller said.

Growers argued that, without the program, prices would be at the whim of the world market, and any savings would be pocketed not by consumers but corporate sweetener users.

"If the sugar price doubles, they will raise the price of a Snickers bar," said Bob Buker of U.S. Sugar. "But if the price of sugar drops, do you really think they are going to lower the price?"

Everglades advocates like George sensed an opportunity.

"George figured it out early that sugar was the economic reason why bad things were happening in the Everglades," Charles Lee of Audubon Florida says, "and why progress to resolve those things couldn't be made."

A FEW WEEKS BEFORE George Barley's *60 Minutes* appearance, Charles Lee arrived at an Orlando airport, a red binder tucked in his bag. The binder contained a report produced by an independent consulting firm. The report was titled "The U.S. Sugar Industry: A Briefing Book."

The briefing book was composed of fifteen chapters and a glossary, and the chapters had names like "The U.S. Sweetener Market," "The Florida Industry," "U.S. Sugar Program Overview," and "Proposals for Change." Charts, graphs, and maps provided further data on industry trends. The first chapter, on the world market, began:

- Sugar is produced in over 100 countries around the world and current annual output is 111 million metric tons. (A metric ton is about 2,205 pounds, so it is about 10 percent bigger than the "short tons" used in the United States.) U.S. production represents about 7 percent of the world total.

- Sugar is produced from sugarcane in the tropics and from sugar beets in the temperate zones of both the Northern and Southern hemispheres. Most statistics are kept in terms of "raw value," i.e. the weight of raw sugar produced by a cane mill, before it is refined into pure white sugar. Raw sugar still has molasses and other impurities coating and imbedded in the sugar crystals.

- The factories that extract sugar from beets produce only a refined white sugar. USDA uses a factor of 1.07 to convert between white and raw sugar since it takes about 1.07 pounds of raw sugar to produce one pound of refined sugar.

- World sugar production can be quite variable due to weather and the cyclical effects of investment in production facilities. Consumption is much more stable and is not very responsive to price changes. Coupled with government policies around the world that seek to isolate national sugar markets from the world market, this can result in wide swings in world prices.

- The world price for raw sugar in most years is no more than half of the U.S. price, which is supported by the government's sugar program.

Lee was part of a plan designed by environmental and sweetener user advocates to hit every major newspaper editorial board in Florida in just two days on the issue of the sugar program. Two teams would execute the tour de force, and each member carried a briefing book. Mary Barley had produced the itinerary. After leaving Orlando, Lee and his team would meet with the *Daytona Beach News-Journal* and then the *News-Press* in Fort Myers. The next day the team would meet with the *St. Petersburg Times* (now the *Tampa Bay Times*), and the *Tampa Tribune* (now defunct). The other team had meetings scheduled with the *Sun Sentinel* in Fort Lauderdale, the *Miami Herald* offices in Miami, and *Fort Lauderdale and Florida Today* in Melbourne.

Lee had been an advocate for Florida's environment since he was fifteen and his mom had driven him to environmental meetings. Now he was in his mid-forties and the primary media spokesman for the Florida Audubon Society, now Audubon Florida. He had a round face, round glasses, a gray mustache, and brown receding hair that faded to gray at his temples.

For the first time, environmental groups like the Florida Audubon Society, prohibited as nonprofits from giving politically, had financial backing on the Everglades—thanks to George and his penny-a-pound campaign—but it was still not enough to compete with Big Sugar's spending.

"The only thing you can do," Lee says, "is work the media and bring the facts to the attention of the legislators that are motivated by listening to facts and bring pressure on the legislators by virtue of grassroots campaigning."

Environmental and sweetener user advocates teamed on conference calls and in meetings in Washington. They sought to break up the cane growers' alliance with beet growers that put Big Sugar in congressional districts across the country. Audubon members in the Midwest, for instance, wrote to their congressional leaders that beet growers were benefiting from the same sugar program as cane growers, who were wreaking havoc in the Everglades.

Lee wrote a letter sent by the Florida Audubon Society to thousands of contributors to then senator Bob Graham, a Democrat.

"We sent them an information packet. It was basically, The Everglades are being destroyed by the sugar industry, and we're trying to get Bob Graham to force them to stop and to help reform the sugar program, and we're asking you to join us by contacting Sen. Graham," he says.

"After the letter dropped and Bob Graham started to get calls—and he did get calls. He got a lot of calls from his campaign contributors—Bob Graham called up whoever was the CEO of Audubon and really started chewing on me for having sent that letter, and I knew I had to fight a backfire within Audubon because Bob Graham was not happy."

Still, Graham did not budge.

"The sugar guys are simply experts at targeting strategic major campaign contributions to candidates individually, to party coffers, and to PACs," Lee says.

Before boarding the plane for his tour of editorial boards Lee had received a memo from George Barley with some thoughts on the scheduled meetings:

> Jim McNair has written business columns defensive of sugar. He wrote one saying penny a pound would put them out of business. He should be approached on economic grounds. Hazen & Sawyer's paper on jobs, which may be ready Monday morning, should be given to him. I'll send him a sugar briefing book. He has in the past relied on USDA economic data on sugar profits. He should be encouraged to talk with Grace Johns at Hazen &

Sawyer. He should also be given Prof. Polopolus's presentation to the Governor's commission on a Sustainable South Florida and a copy of Prof. Boggess's peer critique (working for us). I would press the point, if I was there, that this is a dirty industry, and Florida needs clean industry. This is also an industry that through political influence has made the taxpayers pay $400 million to clean its pollution, money that is needed for other parts of the Everglades restoration. I'd stress the idea of an economic development zone for the EAA, one that will attract clean industry, higher paying, healthy jobs. It is also an industry that doesn't tell the public the truth—just look at the "our water is cleaner than rainwater" ads run several times in Florida Trend by U.S. Sugar. They have also continuously exaggerated their impact on the economy. I'd point out what a small part of the south Florida economy they were, but I'd be sure the number I used was backed up by a source. I'd also try to get the Publisher (Editor) of the Broward Edition of the Herald in the room, he was very excited over penny a pound after I showed my slide show to the Environmental Coalition of Broward County a year ago. Patty Webster knows his name, he sat at the head table.

The memo continued:

Team B [Lee's team] should try to hit more than two newspapers. I'd try and start earlier and add something to this schedule. Maybe the Bradenton Herald, it is in between Daytona and Fort Myers. Or the Naples paper after Ft. Myers. I'll have a plane available at Showalter Airport for all day Tuesday for Team B. It has a seat next to the pilot, two seats behind it, and two with somewhat limited leg room behind that. Please appoint a coordinator for this flight so I can furnish the pilot his number and vice versa.

THE SUGAR PROGRAM guarantees growers a minimum price by controlling supply and limiting imports through tariffs. Growers are quick to point out the program does not represent a subsidy. When the supply is high and the price low the program goes further, allowing for loans that farmers can satisfy by forfeiting their crops to the government. Since the 1970s, as a result of the program, the domestic price of sugar has been nearly double the world price most years. This price difference represents the program's most significant cost to food consumers, according to Michael Wohlgenant, a North Carolina State University agricultural economics professor who has studied the sugar program.

A report Wohlgenant authored estimates that the program costs food consumers $2.4 billion a year and provides sugar producers with $1.4 billion a year in benefits. Other studies support Wohlgenant's findings. A 2000 report from the Government Accountability Office, the most recent available, found the program cost domestic sugar and sweetener users $1.9 billion in 1998. A 2006 US Department of Commerce report found higher domestic sugar prices were a major factor in the loss of more than ten thousand jobs between 1997 and 2002 for manufacturers using sugar in their products. The report estimates nearly three confectionery manufacturing jobs are lost for every sugar-growing job saved through the sugar program.

"It wouldn't be feasible to grow sugar in Florida without the price support," says Lance deHaven-Smith, professor emeritus at Florida State University and author of several books on Florida politics. "It's called a collective action problem. The people who have a stake in it see their interest, and they are all over it. They are constantly working on it. But the other side, the people who are paying a little more for their cereal in the morning, they never see it, and so they never mobilize. . . . That's the problem. You don't have a balancing of the interests, but this is a problem all across the board. It isn't just sugar."

The inflated domestic price of sugar has also contributed to the widespread use of high-fructose corn syrup as a lower-cost alternative. Meanwhile, by limiting imports into the United States, the sugar program has depressed the world price of sugar by some 8.5 percent, Wohlgenant found, affecting the bottom lines of some of the world's poorest farmers.

Precedent exists for ending the program. In 1974, when commodity prices boomed because of an energy crisis, inflation, and global commodity shortages, lawmakers suspended the program.

Even if the program were halted, sugar growing would likely continue in the Everglades Agricultural Area, farmer Rick Roth says. That's because, unlike crops such as lettuce or cabbage that can be plucked and shipped, sugarcane must be refined in a factory, and that limits where it can be farmed. The Everglades Agricultural Area is unique for its sugar-growing infrastructure. The crop likely would endure there because the infrastructure is there.

"It's the crop that is best suited to the weather and the soil type. So sugar cane is going to be the crop of the future," he says. "It's a combination of domestic sugar policy and agronomy that gives sugar the advantage."

For families, farming requires a massive investment spanning generations, he says. His family has spent two generations acquiring Roth Farms' several thousand acres. It's a risky business, dependent on weather and other variables.

"What the sugar program does for me as a farmer, it gives me a stable plan I can go to a banker and say, 'Look, I'm producing this many acres of vegetables and this many of sugar cane,'" he says. "'Will you loan me $2 million to plant my crops?' And they go, 'Yeah.' They want the estimates: What are the costs? And what are you going to sell sugar for? And having a sugar program means the banker knows with certainty I'm not going to sell the sugar below this price. He knows I'm going to get at least 18 cents, and he knows I'm going to sell all of my sugar."

Roth believes that without the sugar program the US price of sugar would become volatile, hurting consumers as grocers hiked prices, but demand would prevent the price of sugar from plunging.

"It's really thanks to you," he says. "It's called more mouths to feed. As you have more people in the world demand is exceeding supply, and that's what's driving the price up."

ELEMENTARY SCHOOL STUDENTS waved signs proclaiming, "American Sugar—420,000 Jobs" and "Save My Daddy's Job." The Glades High School marching band played.

Days after Lee's tour of newspaper editorial boards a congressional subcommittee traveled from Washington, DC, to the Everglades Agricultural Area to hear testimony on the sugar program. Hundreds packed a community college auditorium for the meeting. Many were unable to find seats, and the proceedings were also shown outside on a closed-circuit television.

Local leaders and farm families spoke emotionally of how ending the program would endanger their communities. Nationwide some 420,000 jobs would be lost, they said, 40,000 right here in towns like Belle Glade, Clewiston, Moore Haven, and Pahokee. Some seemed to believe the crop's very survival hung on whether the program endured.

"In my town if a store employing half-a-dozen people moves out or goes out of business we all feel it," Pahokee mayor Ramon Horta said. "I can't imagine the devastation that would occur should the sugar industry dry up and blow away. We go as sugar goes."

A third-grader whose father worked at a refinery explained how important agriculture was to her family and community. A high school principal talked of the stability sugar gave the region and noted that the auditorium itself had been donated by the industry. Florida commissioner of agriculture Bob Crawford described the program as economically and environmentally sound.

"If we didn't have a sugar program I don't know what the price of sugar would be," he said.

Some accused corporate sweetener users of greedily aiming to kill the program so they could buy sugar for a few cents less without passing along savings to consumers. One woman whose family raised sugar and cattle held up a can of Classic Coke and another of the artificially sweetened Diet Coke. Both cans of Coke cost the same, she said.

"If America's sugar policy is dismantled the American consumers aren't going to see a penny in savings," she said. "The cost of sugar has nothing to do with what we all pay for a can of Coke or a Mars Milky Way."

Only one opponent of the program was invited to testify. Karen Lee, a Key Largo mortgage broker, began by saying it took courage to speak where so many disagreed with her position. She then talked of how the program inflated the domestic price of sugar and said the industry was responsible for only a third of 1 percent of the state's gross economic product, a figure provided by George Barley. She drew hisses as she brought up the issue of pollution in

Florida Bay. She expressed sympathy for farmers' fears but said Florida Keys residents felt similarly about the bay.

"Our jobs are at stake today," she said. "It's not just a possibility. It's a reality."

Her testimony had been arranged by George Barley, along with a news conference outside. Fowler West, the lobbyist, had taken care of the details.

"Karen Lee's testimony and the outside press event put a huge dent into the hearing's effectiveness. In fact, we received about as much press coverage as did the hearing itself. George was elated!" West wrote years later. "My conversation with George about our success, made from a Belle Glade pay phone, was the last time I spoke with George."

Eight weeks later George Barley was dead.

9

THE PLANE CRASH

On June 23, 1995, George Barley had a day-long meeting in Jacksonville on the Everglades scheduled with the US Army Corps of Engineers. He and Mary rose early for his flight out of Orlando International Airport, sharing a pot of coffee for breakfast. Then they shared a kiss good-bye, although he would call five or six more times by 9 a.m.

At the airport there was a mix-up with George's flight. He thought it was delayed, but it wasn't. After he wandered away from the gate the plane took off without him. Characteristically undeterred by this inconvenience he chartered a Beechcraft 58 Baron at a general aviation airport near downtown Orlando from a company he had used many times before.

"I'll see you when you get home," Mary told him when he called with his latest update. "He was aggravated, and he wanted someone to talk to."

Soon after taking off, pilot Mark Swade radioed, "We've got an engine failure. We are declaring an emergency." Then he said, "We're going to go down."

The plane soared through an overcast sky above a lake, banked left, and hit a tree. It rolled right, and its right wing hit the ground. The plane cart-wheeled, slid into a tree, and erupted into flames. Just before the crash, witnesses heard the plane's engine misfire and sputter. The duty pilot at the airport would describe the disaster for the Orange County Sheriff's Office this way:

> I observed the twin engine Beechcraft Baron 58 laying [*sic*] in the front yard of the residence located at 5587A Lehigh Avenue just west of Solandra Drive. The plane was facing in a southern direction with the tail coming to rest between two medium size trees. The tail section was twisted partially off and standing somewhat erect and both wings were embedded around the trunks of the trees. Furthermore, both engines had been torn from each wing and pieces of the plane were lying around the ground and in the trees. The entire plane was burned along with both victims who were still inside the aircraft. Looking back to the south, evidence of the path that the aircraft took to reach the final resting site could be determined.
>
> The aircraft apparently flew through a large oak tree located in the yard of the residence located at 912 N. Solandra Drive. When this occurred, the aircraft sheared of[f] the entire top of the oak tree causing branches to rain down on the yard below. Also in the debris were various size parts of the clear plastic type material that had apparently came from the aircraft. The aircraft continued traveling in a northerly direction passing the following yards and leaving debris in them: 918, 1002, 1008, 1014 N. Solandra Drive. At the yard of 1014 N. Solandra Drive, the aircraft struck a 4 foot high chain link fence with possibly a wing tip. From this point northward, a furrow was dug into the ground by again possibly one of the wings suggesting a cartwheeling action of the aircraft. The aircraft then struck the power line running east/west on the south side of Lehigh Avenue causing wire(s) to wrap around the left engine and with enough force to move the power pole several inches out of place. The aircraft continued across the roadway and eventually came to rest tail first after striking two trees with the

wings. The aircraft then caught fire and burned profusely until being extinguished by the Fire Department. The tail section of the aircraft was twisted around the fuselage, and pieces of the wings and other aircraft parts were suspended in the trees several feet in the air. The force of the impact caused the engines to wrap around the tree bases. The propeller tips of the right engine were twisted and bent backwards while one complete propeller blade of the left engine was still untouched and in the featured mode. Scattered on the ground from the impact point with the chain link fence to the final crash site were many various size pieces of the aircraft. Both victims were still inside the aircraft which came to rest facing in a southerly direction.

From the tree in the rear yard of 912 Solandra Drive to the chain link fence located on the north side yard of 1004 Solandra Drive, the distance was measured at 267 feet. From the chain link fence to the final crash site in front of 5587A Lehigh Avenue, the distance was measured at 108 feet.

The plane was a Beechcraft Baron twin engine, model 58, serial number TH809 approximately a 1978, operated as a lease back charter airplane by Air Orlando. Victim # 1, Swade, was the pilot in command of the aircraft and was employed by Air Orlando as their chief pilot. The passenger, Victim #2 Barley, chartered the aircraft after missing his earlier commercial flight from OIA to Jacksonville, FL to attend a meeting there. When the charter flight was booked, Victim #1 Swade went to the Air Orlando hangar and told the chief mechanic (Witness #1 Melde) that he needed the Baron out right away for the flight to Jacksonville. The aircraft was then readied for the flight and the plan was taxied away by Victim #1 Swade to pick up the passenger Victim #2 Barley.

According to several witnesses on Lake Barton, the aircraft's left engine was not operating while taking off or just after lifting off from Orlando Executive Airport runway 07. Furthermore, the aircraft was having trouble gaining altitude

> and barely missed the tops of trees and buildings as it turned in a northerly (left turn) direction. The aircraft then disappeared from sight and then a large fireball and smoke was visible above the trees and buildings.
>
> Other witnesses located in various locations around the crash site and along the route that the aircraft took on its short flight describe roughly the same thing in their sworn statements. The plane had trouble gaining altitude after lifting off from the airport; some type of unknown problem(s) with apparently the left engine; hearing strange noises coming from the aircrafts engines; observing the aircraft strike the large oak tree and ground; and then observing the fireball and smoke. The witnesses describe the fire as intense and that there was nothing they could have done to help either of the victims inside of the aircraft.

Pilot Mark Swade was thirty-two; George Barley was sixty-one.

AT JACKSONVILLE INTERNATIONAL AIRPORT Lewis Hornung waited.

George's scheduled meeting was with Hornung, who managed the US Army Corps of Engineers' Central and South Florida water projects, and Colonel Terry Rice, the corps's newly appointed top engineer in Florida. No doubt George wanted to talk about Florida Bay.

By then Florida's booming population was exerting growing pressure on the Everglades' lifeblood water, an essential drinking water source that had been drained to a fifth of its historic flow. A water shortage was projected by 2010, improbably in a state showered with a whopping 55 inches of rainfall annually. Declining habitats threatened native fish and wildlife like the iconic alligator, but no one agreed on how to fix the problems or who should pay for the work.

To sort all of this out Congress in 1992, as part of the Water Resources Development Act, had ordered the Central and Southern Florida Project Comprehensive Review Study, or "restudy," as everyone called it, of the mid-century effort that had drained the Everglades and made way for modern Florida. Charged with the restudy was the army corps, the same agency that had brought the watershed to the brink in the first place. For the engineers, with a vast history of contributing to environmental degradation, the restudy represented the start of a new era. They now wanted to be "green engineers" of the future, known for fixing ecological problems. The Everglades represented their first restoration project.

"It was very controversial," Hornung says. "We were setting a precedent."

George was prepared to speak on behalf of his beloved Florida Bay, to explain that what the bay needed most was more freshwater and that the best way to provide that was to restore a more natural flow in the Everglades. He carried with him materials on the river of grass's hydrology and the sugar briefing book.

"He was very motivated to come talk to us," Hornung says. "I spoke to him by phone once he had his flight canceled. He told me he was chartering a plane, and he would be there, and I agreed to meet him."

But George never showed. Hornung called his office, but no one knew where he was. Hornung waited some more and called again. That was when he learned of the plane crash.

AFTER KISSING GEORGE GOOD-BYE Mary had left for a seminar to hone her computer skills. Soon after arriving she got another call—but not from George. Someone from the airport where George had taken off was calling to say there had been an accident and that she needed to come right home.

At home Mary discovered seated around her dining room table an airport executive, law enforcement officer, and several other people. A housekeeper had let them in. These were the people who delivered the terrible news that her husband was dead, that he had perished in a plane crash along with the pilot, the plane's only other occupant, and that no one knew yet what had caused the tragedy.

"I'm sure I just about didn't believe it," Mary says. "It was a long, long time before I accepted the fact that he was never going to come back home."

DAYS LATER the Orange County Sheriff's Office report on the crash contained the following chilling information: "The day of the accident, a fax was received from [the Greater Orlando Aviation Authority] from an unknown source stating that the plane should be checked for possible sabotage. No further information about the sender or his motives is known. As stated in the report, the flight was unscheduled and the plane was pulled out of the hangar by the chief mechanic and pilot just prior to the flight taking off."

No evidence of foul play was ever found. After a four-month inquiry, investigators concluded that a plane defect had caused its engine to fail, and that Swade responded erroneously by extending the landing gear rather than raising it. Swade, a married father of four sons with no life insurance, managed to steer clear, though, of apartments, homes, and a packed day care center. The National Transportation Safety Board (NTSB) report read:

> Shortly after takeoff on rwy 7, the pilot reported to FAA controllers he had an engine failure. The controller observed the landing gear retract at this time and then immediately extend. The witnesses observed the acft [aircraft] at tree top level making a turn to the north with the landing gear down and the left engine shutdown and the propeller feathered. The acft descended and struck the top of a tree with the wings level. The acft then rolled to the right 90 degrees and struck the ground. The acft then cartwheeled and slid backwards into a tree where it came to rest. Examination of the left engine showed the engine driven fuel pump drive coupling was rounded off and not turning the pump. The drive coupling was too short for the model fuel pump drive shaft. A 1993 FAA airworthiness directive that calls for checking for the proper length drive coupling had not been complied with. The flight manual for the acft states that the first action to be performed by the pilot after engine failure is to raise the landing gear. The manual also states that single engine performance is predicted on the landing gear being retracted.

The report stated further: "The National Transportation Safety Board determines the probable cause(s) of this accident as follows. The failure of the pilot-in-command to insure the landing gear was retracted following loss of power in one engine resulting in the aircraft being unable to climb and clear trees in the aircraft's flight path."

Years later I was able to reach the NTSB's lead investigator on the crash, Jeffrey Kennedy, who emphasized that George had changed his plans at the last minute, leaving little opportunity for anyone to tamper with the plane. For some, though, the investigators' conclusions did not settle fears.

"People who worked around George said they went in and locked their doors," says Mary, who hired attorneys to conduct their own investigation, but they turned up nothing.

Adds Kevin Barley, George's nephew: "I remember immediately getting in my car. I drove out there and looked at the plane on the ground. I mean, it was terrible, and we kind of all assumed he must have gotten taken out. We wanted to pin it on someone."

George's daughters worried they were under surveillance.

"We were kind of creeped out because after he died it just seems like we were being followed. We just felt like there were cars following us, but it may have been our imaginations," says Lauren Barley, George's oldest daughter. "We would see these vans on the 520 coming into Cocoa Beach. We felt creeped out. We felt like people were closely watching us. We just felt like we saw cars, the same car. We felt weird about a lot of things."

Mary has a story to tell about a scuba-diving trip she and George took in the South Pacific not long before his death. On their cruise ship, she says, they met a couple from New Jersey. The man worked in trucking.

"That tells you right there what he was doing," she says. "He was a mafia guy. You just don't have what he had in the trucking industry in New Jersey unless you're part of the business."

She says one evening the two couples were seated together for dinner, and eventually the conversation got into the Everglades and Big Sugar.

"You need to really watch out for these guys," she says the man told George. "They're really, you know, bad guys, and you have to really be careful."

George abruptly ended the conversation. "Don't talk like that in front of my wife," he said.

Mary says: "That was when I thought, There's probably more going on here than I know about. . . .

"I know mafia people that can do things that you'll never find out, so there's always just that one little tweaky 1 percent that says, 'Well, maybe they did get away with it,'" she says. "But on the whole after listening to all the people that we sued, because what you do, you sue everybody, and they've got to go out and figure out what happened, and then we had a big mediation. And I think could they have? Yes. But likely? Mmm, like a 1 percent chance."

AFTER CHARTERING the doomed plane, George called Charles Lee of the Florida Audubon Society. Lee had helped George prepare for the meeting, and George wanted to offer Lee a seat on the flight one last time, but Lee had a schedule conflict. Later that afternoon, after learning of the crash, Lee dialed into his voice mail.

"There was one [message] from George, left days earlier. He had only been able to attend part of an Everglades Coalition meeting in Ft. Lauderdale," Lee would later write. "His words were as crisp and clear as if he had been right on the line.

"'This is George Barley. Sorry I had to leave early. Give me a call and let me know how it went.'"

10

TOLERATION AND PROCESS

"Let me put it this way," Mary Barley says. "In three days I lost 20 pounds. Don't even ask me how you can possibly do that, but I did. I recall eating, absolutely. I think it's just how stress can get to you. I went from 120 pounds to 100 pounds, and then that day, it's just, I don't know. I'm sure it's back in my mind somewhere. Maybe as we talk through it. I can see me sitting at the table."

After learning her husband was dead, Mary left her dining room table and the somber group seated around it and began calling people. The activity kept her occupied, helped her keep it together. First she called George Barley's friend Paul Tudor Jones.

"It can't be," he said. "You're sure? You're positive? I don't believe it. Where is he? What happened?"

"I don't know," she replied. "All I know is that they're telling me he was killed."

Then her home filled with a "sea of people, just wave after wave, and 'You have to do this and you have to do that. Here, we want you to eat,' and you don't want to eat. People are at you constantly, caring about you, but you're wanting to say, you know, 'Just leave me alone. I don't want to deal with this.' "

Her heart ached, hurt physically.

"You're just kind of like in a haze, you know, and amazed. You're getting through there, but you're not sure how you're making it. And you just keep turning the next corner, and somebody else is asking you a question or coming over, and people I knew and people I didn't know, reporters," she says. "There was always somebody wanting you to do something else, making a new decision or what you are going to do and la-dee-dah, and I don't even know what I was doing. I was just playing a role I think.

"I'm not 100 percent sure. It's almost like an out-of-body experience, when you're like reacting. One part of your brain is saying, 'This is what you have to do,' and the other part of your brain is totally shut down, the emotional side that just says, 'I can't deal with this, so I'll go and do that and that way I can function.' . . .

"I can remember trying to go to sleep that night, and you're just, you can't. Your mind just won't turn off, and I remember they didn't go that day. They went the next day, to identify the body," she says. "Because I remember almost halfway falling asleep [and] jumping up like at 4 or 3 o'clock in the morning, and somebody was telling me, like, 'Get me out of this place.' Like in my head. Like George was saying, 'Get me out of here,' and I mean I know what it was. He was in a cooler or someplace, you know, Lord knows where they had his body in a bag, and I remember thinking, 'Oh my gosh, I've got to get down there and do something about it.' "

AT THE TIME OF George Barley's death he was at the forefront of the Everglades movement, his stature rivaling that of Interior Secretary Bruce Babbitt or Senator Bob Graham. Few since Marjory Stoneman Douglas had administered such a dose of attention and intensity.

He was a leader in two campaigns: one to get a penny-a-pound fee on sugar on the state ballot and a second to eliminate a federal price support program for the industry. Less than two weeks after his death he was scheduled to be in Washington on the sugar program.

"He died in the line of duty," Paul Tudor Jones said.

Days after George's death some 250 people turned out for a memorial service at an Orlando public garden. Photos showed the outdoorsman fishing in the Florida Keys, grinning as he held up a grouse in Scotland. A display of mementos included a lollypop with a "Lick Big Sugar" wrapper. A bagpiper played. Mary, wearing a black dress, smiled for a photographer.

"It was the hottest day. Oh my god. It was absolutely stifling. There wasn't a breeze. There was nothing," she recalls. "We were all celebrating. It wasn't a morbid thing. It was a good time."

After everyone took their seats, George's friend Tom Lewis spoke first, drawing laughs:

> Welcome. I'm Tom Lewis. All of us who knew George knew that he had a knack for putting us in a hot spot, and he did it today.
>
> I considered it my great fortune to have been a friend of George's for almost twenty-five years, and as you look around, this is just George's sort of day: Outdoors among his friends with his wife and his family. Beautiful women. Sky of blue. George would have loved today.
>
> George was the ultimate connoisseur of life. Even more he was a man who made a difference. This morning George's family and his close friends will join with me in remembering this difference. In the two decades I have known George his passion and zest for life have never ceased to amaze me, and I am sure all of you.
>
> Dinner to George wasn't a meal. It was an event, a lifelong quest to find the very best—and hopefully to find one of his friends to pick up the bill. Wine was not a beverage but the elixir of life that led him from bottle shop, to vineyard, and ultimately to the wine cellars of his friends Tom Dittmer and Paul Jones.
>
> Olive oil. Last year George decided that olive oil was a worldwide significant issue. That set off an endless quest for the very best.
>
> Hunting. I know all of you have your favorite George Barley hunting story. Mine was a Maryland duck hunt hosted by his friend Paul Jones, who thoughtfully provided golf carts to take

us all to the duck blinds. This courtesy became a road race putting George's driving skills against the world-renowned champion Jackie Stewart. Guess who won?

In the most recent years, the Everglades, which has so captured George Barley, has taken his passion and he became a crusader. George was an action man whose credo was, Nothing in moderation. When George got involved in anything it was flat-out all the way. It was that great passion, his humor, and his joy that he has brought to all of us and that we are all so blessed to have known him and known of. George, you are my hero, and I hope you're up there watching us.

Another friend, David Davies, spoke next:

My name is David Davies, and I'm from England and Ireland and I've been asked to say a few words on behalf of the legion of George's friends overseas. John Masefield, the English poet laureate, in the middle part of this century wrote a few lines in the dark days of the Second World War. His lines were, "But oh, my dear, how rich and rare and root down deep and wild and sweet it is to laugh." And how those lines encapsulate the memory of our friend George Barley, who we all love and adore.

Who can forget that infectious, mischievous chuckle, which made each one of us a conspirator with George in his pursuit of life. Who can forget that sense of humor, which could always find the one ray of sun in a dark sky. Indeed, how could anybody but George have the extraordinary nickname of Marmalade, which my overseas friends bestowed on him eleven years ago in Hong Kong. And for those of you who don't know the background of that story let me say it has as much to do with a bottle of cognac as it does with a jar of Robertson's Golden Shred Marmalade.

Under that fun-loving exterior was a man of integrity and passion who would always fight for what he believed in. You know better than I do his crusade for the Everglades, but how he could infuse all those who came near him with his deep concerns. I saw it myself at first hand when, inspired by George, our own Prince of Wales became most interested in the whole project, and I was able to send George just a few months ago a copy of his own, the prince's own, handwritten four-page letter to Vice President Gore.

Finally I would like to pay tribute to George's zest for life, to see new places, meet new people, always inquiring, always sensitive to others. I am going to miss those endless faxes I seem to receive every month. Quote, "I've been in a remote area of Canada for eight days and now back to the paperwork." Quote, "Let's plot something else. For some reason we are dying to go to Vienna for Christmas and to the South of France for the summer." Quote, "We plan to go to Mexico in mid-July to see our friends David and Sara." Quote, "We've just reviewed a video of Zambia. How wild and remote. How I've longed to go there." And I remember so well the last trip that my wife, Linda, and I, together with George and Mary, made to Cambodia to visit the amazing temples of Angkor Wat. There is no better traveling companion ever.

Then George's daughters spoke:

Hello. I'm Lauren, George's oldest daughter. This is my sister Kathy and Mary, and we would like to say a few words about our father, and the one thing that I would like to say that I will always remember about my dad is his commitment to excellence.

There were never any excuses. No excuse was good enough. You got the job done, and you did it right, but you also did it with love for the task and those around you. And we are going to miss our father, and we love his friends dearly. We love Mary, and we are going to try to continue on the great task he set before us, and that is to walk in excellence, achieve in our lives, do it to the very best.

> My father did everything to the best: cooking, photography, dressing, his home, his things, his friends, his hunting. Everything he did, everything he touched, he did it first-class. We love you, Dad. We are going to miss you.

Kathy followed her sister:

> Last night I wrote a poem for my father, and I'd like to read it and share it with you all today. And it's entitled, "Farewell to My Hero."
>
> *Father, we shall miss thee, the one who so loved the sea.*
> *But now it shall be that you are much more than memory.*
> *You have made your mark in life. God knows it was not in strife.*
> *For you have left your mark so that we shall not live in the dark.*
> *Great men are far and few. Too bad there weren't more of you.*
> *Your talent is hard to comprehend. Your being was a godsend.*
> *We could all learn a thing from you, not brave enough to emulate*
> *what you do.*
> *You lived your life with such zest. That's what separated you from*
> *the rest.*
> *You would never want us to cry or exclaim, "Oh, why?"*
> *You knew what you had to do because that was just you.*
> *Your legacy will live, for you had so much to give.*
> *Our hearts shall ache. The loss of our true hero is hard to take.*
> *We all love and miss you so. Too bad you had to go so soon.*
> *It's only June.*
>
> Thank you.

George's youngest, Mary, was the last of his daughters to speak:

> In the past few days there have been many references to my father, George Barley, as being the developer and George Barley

> the environmentalist. But no one has mentioned his greatest accomplishment yet, and that was his role as a father.
>
> My father had very high standards for himself and for everyone in his life, including myself and my two sisters. As teenagers we balked at his demands for excellence, but as adults we are all grateful and will strive to maintain those standards throughout our lives.
>
> Tomorrow I will be celebrating my thirty-second birthday, and the gift my father has given to me for the rest of my life was his involvement in the direction of my life throughout. He taught me so many things and opened so many doors of opportunity for me. I will be forever grateful.
>
> I hope his very many wonderful qualities—his keen intelligence, his tremendous enthusiasm for life, and his great capacity for laughter—will continue on with all of us and with our children. I feel very blessed and proud to have had him as my father, and I will strive to continue to make him proud of me.

A pianist performed "Amazing Grace." Next to speak was Nathaniel Reed, a highly regarded leader in Florida's environmental movement who had served as an undersecretary in the US Department of the Interior and chair of the South Florida Water Management District board:

> I'm Nathaniel Reed, a fellow fisherman, an unabashed environmentalist who loves this state, as George did.
>
> George Barley was a man of unusual brilliance, determination, humor, resilience, and unlimited energy. When George saw his beloved bay disintegrate before his very eyes he reacted with courage and ethical conviction that man does not have the right to destroy what God has created.
>
> It did not take George long to realize that Florida Bay was connected to the greater Everglades ecosystem, and that insensitive engineering and gross mismanagement of water were the dreaded combination that have brought Florida Bay to near

> death. Initial anger, initial finger pointing, well-deserved, then passed, and George reassessed the situation and his role.
>
> That process of maturation developed him into one of the most formidable environmentalists in Florida history. With his usual enthusiasm, his aggressive can-do attitude, George plunged into the fray. Supported by Mary, his love of life, and you, Paul, Curt, fellow friends, and supporters, he began to make things happen at every level of government. George became an environmentalist in the truest sense of the word, understanding Aldo Leopold's vital message on connections: Florida Bay is connected to the greater Everglades system, and to improve Florida Bay we must improve the entire system.
>
> I have lost my bonny friend, a friend for all seasons. Florida has lost a fearless champion, a powerful voice, a powerful force for public good. Death be not proud. You have stolen a great man from his wife, family, friends, and his beloved state. George, we collectively vow to continue your work with renewed vigor and determination. And may our paths cross again, dear friend. And until then, rest in peace. God bless George. God bless my shining knight. And thank you for all the joy you have brought to our collective lives.

Another friend, Tom Dittmer, followed:

> My name is Tom Dittmer. I am one of George Barley's friends. George is my teacher and my mentor of fun.
>
> I grew up in Iowa. I thought that hunting and shooting was walking down a corn field in Iowa. I didn't know that to be a true gentleman you had to go to Europe twice a year to shoot. Then I thought a matching pair of guns was having two Remington 1100s. George informed me that I needed side-by-side Purdeys. Then he also informed me it had to be Italian, Spanish, and English. George thought my Blue Nun, Mateus, and Liebfraumilch was not quite up to par. He took immediate charge

of my wine cellar. I have a small mortgage on my house now, but it will be paid off soon.

I'm from Iowa, and in Iowa a swamp is something you fill up. That's where the snakes are and the bugs, and it was what I really thought. That wasn't the definition George had. He slowly convinced me that the swamp was the heart of the universe, and he had to save it. I finally won one, though, and that was in fishing. He tried to get me to love fishing, and I resisted, and it was very hard because he's, as you know, very passionate.

As you can see by all these things, George wanted you to love and have the passion for what he loved, and he put a lot of work into it. A lot of time and effort. The only problem with it was that you had to quit your job and have a very large trust fund.

If life is measured by love or passion, then George lived life. If life is measured by friends you touch and causes you champion and your family and your dedication to make the world a better place, George lived.

Here Dittmer lost his composure. Last to speak was Paul Tudor Jones:

My name is Paul Jones. I'm from Memphis. I was walking to my bedroom one night last fall in my hunt camp in Maryland when I heard laughing coming from the den, so intense you knew somebody's sides were splitting. So I wandered in there, and who was there all by himself but the Harvard-educated connoisseur of fine wine and aromatic cigars, great food, lover of tradition, ceremony, pop history, everything that was elegant in life, and he was watching a movie.

And in between the maniacal laugh of his that we all know and understand, tears were rolling down his cheek. And he looked up at me, and he could barely choke out, "Greatest movie I've ever seen!" And I reminded George that *Animal House* had not won any Academy Awards the year it was released, but that didn't seem to have that much of an impact on his cinematic judgment.

Now, his surprising love of that movie is illustrative of why his friendship was so unique. He was a paradox. He was the ultimate paradox because being a friend of George's was a challenge. You never knew what new adventure that you were going to be taken on. You never knew what to expect from him except that you knew there were going to be 50,000 watts behind it. As a friend, he gave the fiercest loyalty that I have ever experienced. And that's why his absence to all of us is going to be so tough because his life force was so powerful.

Now, as much as I loved George for being a great man—a brother to me, a father to me, at times I felt like he was my son—it's impossible for me to separate him from the cause because his commitment to Everglades restoration made so much context in the sense of who he was, of all his passions: The transfiguring beauty of a sunset on Florida Bay. The vitality of a bird rookery at sunrise. The serenity of a seagrass flat under a noonday sun. Or even your breath as deafening. Those were the beauties of life that George loved so much. He fought in his life for elegance and art, and his love of the Everglades was for that love of elegance, art and art as he saw it, most beautifully and wondrously expressed, which was in the Everglades.

And he wasn't just fighting for himself. He wasn't just fighting for Mary and his daughters, who clearly he was fighting for. He was fighting for people he didn't even know. And why? Why, because for a person that we all know could give no quarter at any point in his life—not his personal life, not his professional life—it was the ultimate balancing act. It was the ultimate form of giving, the ultimate way that he could pursue that ideal of nobility that was so important to every aspect of his life.

Now, in the final scene in the movie *Glory*, there is a scene where the commander of the troops knows that he is getting ready to go into battle in which he is going to die, and he looks at his troops and walks up to the gentleman holding the flag and says, "If this man falls who will pick up the flag?" If George

> had known he was going to die he would have summoned all of us here together today, exactly like this, and he would have said, "If I die, who will pick up the flag?" I see the faces of many flag bearers in this audience, and I swear on your behalf, my personal oath, in George's name, that with the sharp point you bequeath us, we will win this fight.

It was so hot, the kind of Florida heat that smothers like a moist heavy blanket. Suddenly Mary felt aware of George's presence.

"Just for like two or five seconds a breeze came through," she recalls, "just like he was going by saying good-bye."

Remembering this she begins to weep, but it's only for a minute.

GEORGE WAS CREMATED. Mary and his daughters each kept the ashes.

"I tried to put a little in most of his favorite places as I traveled, or places I knew he wanted to visit," Mary says. "The rest will be spread in Barley Basin in Florida Bay, along with my ashes."

For months letters poured in from friends like Nathaniel Reed and Tom Weis, environmental groups, congressional leaders like Senator Bob Graham. Mary also received letters from Jeb Bush, who years later would be elected Florida's governor, and a deputy private secretary to Prince Charles, who wrote that the Prince of Wales was "so sad" to learn of George's death.

Also among the letters was one from David Weiman, a lobbyist who had worked with George. It read in part:

> Working for George, as has been described by many, was a challenge in the extreme.
>
> George was a successful businessman. Along the way, he chose to do what few others—less than one in a thousand—elect to do. Give something back to the community that was so good to him. He did.
>
> All of Florida benefitted.
>
> His passions for "his" Florida Bay were unmatched.

> In many ways, and this may seem odd, his very success in business may have conflicted with his efforts to get Government to "do the right thing"—at least in one regard. George was an entrepreneur. The skills and attributes that make or create business success are almost the opposite of those required in government policy. The very nature of government—at least one I used to know—is based on the principle of toleration. Our system of government tolerates different views. It tolerates divergent opinions. Well, it's supposed to do so.
>
> The act of toleration compels a "process."
>
> Government is about process. It ain't always fair . . . but there's always a process.
>
> That process drove George to utter distraction. It went against every grain of knowledge in his being. Government both operates AND is changed by a political "process." Many people find that disdainful. Put more bluntly, George rejected it. Yeah, he wanted, no demanded the change, but he suffered that process AND that process suffered him.
>
> There's an old saying, the source of which I should know, that goes something like this—our system of government stinks, but it's better than any other on earth. A quarter century of experience in the business gives me a sense of truth to the statement.

George's goal was the same as the government's, theoretically: policy that begins with constituents rather than special interests. He believed sugar growers were not paying their fair share toward Everglades cleanup and instead were corrupting the very elected leaders charged with protecting Florida's environment, which was essential to the state's growth- and tourism-based economy and also residents' health and well-being. He gave his time, money, and life toward getting the government to "do the right thing" for the people, but he was not "the people" in a populist sense. He was a millionaire, a developer, and a globe-trotting fisherman and hunter.

"One would wish that the Congress and the great federal government would act more wisely, but the fact of the matter is it doesn't. It takes individuals who care deeply to motivate the great federal-state bureaucracy," Reed told me not long before his death in 2018. "What is required is total commitment exemplified by Mary and George Barley and the associates that they selected to be their partners in the monumental effort to restore the Everglades ecosystem."

George would not have wanted sadness but anger. He would not have wanted talk but action. Paul Tudor Jones had asked George's family and closest friends, "Who will pick up the flag?"

Mary would.

11

THE CAMPAIGN RESUMES

After her husband's death Mary Barley plunged into what would be the most expensive political campaign up to that time in Florida history. She began by keeping the appointment George had missed with the US Army Corps of Engineers when his plane crashed.

"They made a point to say this was the briefing they were prepared to give George that day," says Clay Henderson, then the president of the Florida Audubon Society, who went with her.

"She had some talking points. She was very firm about how we had the need for restoration to be a priority rather than agriculture," he says. "This was step one into kind of filling the shoes that George had—meeting these people, trying to understand where they were coming from, and trying to advocate a position on behalf of the Everglades."

Mary was grief stricken and inexperienced politically, but not intimidated.

"I am not afraid of anybody and especially anybody in government, and I don't have any problem saying what I feel or what I think should be done," she says. "It's not that I can't be convinced. Maybe I've got the wrong idea, but I'm not just going to accept it because you say so."

HERE IS HOW MARY remembers this time.

"I think I was in shock. I had so much other stuff on my plate," she says. "It was Paul [Tudor Jones] who was saying, 'Who's going to pick up the flag?' And then Paul and I decided I was going to pick up the flag, really. I mean he couldn't do it. He had a business, and he had a lot of responsibility. And it was just a matter of, 'Do you think you can do it?' You know, talking about what you need to do, how are you going to get it done. So we went for it. We didn't necessarily know if it was going to work out. We just figured we'd give it a try."

Mary Barley replaced her husband as head of the Everglades Foundation, the nonprofit organization he had established in 1993, and the Everglades Trust, the lobbying effort launched a year later. She reached out to elected leaders and environmental groups. A few weeks after she met with the army corps, the trust was the beneficiary of a $1,000-a-person fishing tournament in Florida Bay hosted by former president George H. W. Bush. Mary addressed the former president and other anglers during a tournament dinner, lamenting the Everglades Forever Act.

"It was a bill written by the sugar industry, for the sugar industry, and of the sugar industry," she said. "It was a step backward for water quality standards in this state and a prime example of how our political system can be corrupted by motivated wealthy special interests."

Looking back Mary says, "I had no media campaign experience. I didn't know anything about it. It just wasn't my deal, so I never had anything to do with that. So I learned what I did from George, just being around him and around the meetings, and at that time it wasn't emails, it was faxes and telephone calls. And then you would go out to dinner with people and hear all the conversations, and so that's how I learned about the Everglades. . . .

"And then you've got to remember I was also the personal representative on his estate, so I had to right away . . . go and hire attorneys. . . . All I did was work. I didn't go to bed until maybe 2 o'clock in the morning, and then I'd be up early and do it all over again. The phone would be ringing off the hook between the attorneys and the campaign. . . .

"I married the Everglades."

BY THE END OF 1995 Mary had joined in the bitter battle in Congress over the farm bill and sugar program. She also resurrected George's bid for a tax on growers. The turmoil in the toss-up state of Florida leading up to the 1996 presidential election would thrust the Everglades into the race between President Bill Clinton and Senator Bob Dole, making the watershed's restoration a national issue.

While Congress toiled over the sugar program, the coalition of environmental groups and sweetener users now led by Mary Barley began touting a 2-cent-a-pound federal tax on sugar, which drew the attention of Senator Richard Lugar of Indiana. Lugar, chair of the Senate Agriculture Committee, was running for the Republican presidential nomination, and he introduced legislation that would have initiated the tax to raise money for the Everglades. Meanwhile Governor Lawton Chiles and Florida senators Bob Graham and Connie Mack proposed a sugar fee that would have raised a fraction of the funds but that they saw as compromise.

Daily ads in Florida newspapers and on radio and TV stations warned of forty thousand grower-related jobs on the line or, conversely, accused growers of using campaign contributions to buy votes that cost consumers and harmed the Everglades. Some ads described environmentalists as "extremists" whose tax would hurt Florida businesses.

"And who knows what's next," one newspaper ad said. "Special taxes on golf courses? Special taxes on toilets that use water? On paper towels? You name it, they'll eventually tax it."

Other propaganda made an issue out of the Fanjul brothers' status as non–US citizens.

"Aliens Earn Billions In Gov't Bonanza!" a front-page banner headline proclaimed from a newspaper created by sugar opponents and distributed on Capitol Hill.

Then there was an ad printed in South Florida Spanish-language newspapers that featured a photo of Cuban leader Fidel Castro alongside that of Nathaniel Reed, then a member of the South Florida Water Management District governing board, who had helped former Republican governor Claude Kirk establish the state's first environmental protection department.

"What Do These Two Have In Common?" the headline asked in Spanish. Captions suggested Reed wanted to confiscate Florida sugar farms and destroy the industry in the same way Castro had done in Cuba.

"And when they achieve the destruction of Florida's sugar industry, Castro's sugar market will grow, helping him to continue in power," the ad said. "Tell your senators and representatives to vote against this discriminatory tax on Florida—and vote against Castro."

"It was a real campaign," Henderson says. "We knew right away that we had a tiger by the tail."

Senator Bob Dole, then the Senate Republican leader and eager to compete with Lugar for the party's presidential nomination, got $200 million for the Everglades into the chamber's farm bill, but rival President Bill Clinton would not be outdone. Not since Jimmy Carter in 1976 had a Democratic presidential candidate won Florida, a crucial swing state with twenty-five electoral votes at the time. In February 1996 Vice President Al Gore traveled to Everglades National Park to call for $1.5 billion over seven years for the river of grass *plus* a penny a pound on sugar.

A summary of the bill insisted the Clinton administration would "work to ensure that Florida's sugarcane industry contributes its fair share of the costs of the restoration effort, in view of the industry's impact upon the environment." The federal and state governments were to partner in retooling the canals and levees holding back the Everglades' water and acquire some 100,000 acres, farmland mostly, for filter marshes. The administration characterized the plan as the nation's largest environmental restoration ever, surpassing efforts in the Chesapeake Bay and Great Lakes.

Sugar growers shut down their harvest and refining operations the day before Gore's announcement so that some two thousand workers could protest outside a Miami-area country club where the vice president spent the afternoon. Arguing that the plan was politically motivated, the farmers pointed out that they already were paying under the Everglades Forever Act to help clean up the river of grass. One protester wore a T-shirt proclaiming "Everglades Farmers, Endangered Species."

The plan certainly appeared politically motivated. Neither Chiles nor Graham, both Democrats and Everglades advocates, were included in its drafting. Mack, who as a member of the Senate Appropriations Committee would be key in getting the plan through the Republican-controlled Congress, also was not included.

A few weeks later Congress reached an agreement on a farm bill that the *New York Times* described as "the most sweeping agricultural bill since the Depression." The measure phased out price supports for corn, cotton, rice, and wheat, but a narrow House vote left the sugar program intact. The farm bill also contained $200 million for the Everglades. Mary and other Florida business and environmental leaders met with Clinton later that month during a campaign stop at the Biltmore Hotel in Miami. Two blocks away nearly one thousand sugar supporters chanted, "Save our jobs." At one point during the meeting the president smiled for a photographer and held up a T-shirt that read "MAKE THE POLLUTER PAY!" On the couch beside him Mary smiled.

"He is definitely on the Everglades' side," she told a newspaper afterward. "We wanted him to see that we have broad-based support, and that it's time for the polluters to pay their fair share."

Mary Barley with President Bill Clinton at the Biltmore Hotel in Miami in 1996, shortly after passage of the omnibus farm bill

Courtesy of Mary Barley

Mary Barley later got a thank-you note from the president. "My Administration is determined to pursue a comprehensive plan to restore the Everglades and the South Florida ecosystem," Clinton wrote.

The $200 million at the time was the single largest expenditure ever on the Everglades. The money would be used a year later to buy out Talisman Sugar Corp. and acquire 63,000 acres, but more money was needed. Some $2 billion in authorized projects awaited funding. Restoration of the Kissimmee River alone was projected to cost some $500 million, but only $13.6 million was appropriated.

Mary felt sugar growers were not paying their fair share. The most expensive political campaign at the time in Florida history would not be over a public office but the Everglades. For the first time voters would face the question, How much should polluters pay for their own cleanup?

MARY WAS MIDWAY THROUGH her evolution from grieving widow to leading Florida environmental advocate when in March 1996 she received the following memo from lobbyist Fowler West:

> Mary, you have become a fine public speaker who makes an impressive and convincing presentation. You do not need media training in the area of delivery.
>
> My only suggestions are these:
>
> 1 Use a lectern that is not too high. You are not a tall person. Find a lectern not much higher than your waist.
>
> 2 Focus on projecting confidence—NOT ANGER. I know you are angry, but look at the fool [U.S. Sugar Corp.'s Bob] Buker makes of himself by reeking anger. You have the right position, so be CONFIDENT.
>
> You will do a great job. I have been asked to go over your remarks ahead of time, though I am sure I will have few, if any suggestions.

I HAVE BEEN AROUND SPEECH MAKERS FOR WAY TOO MANY YEARS, AND I WOULD PLACE YOU IN A VERY HIGH CATEGORY. YOU WILL BE A HIT!

IN APRIL 1996 Mary Barley spoke at a wetlands seminar at the US Department of State in Washington. Her prepared remarks read in part:

> I have asked myself many times what enabled George to keep up the fight even though we suffered setbacks in Congress, in the state legislature, and even on the ballot initiative. What led his friends to take up his charge when he left us, undoubtedly was his fierce dedication, his ability to lead, and his vision itself. George led by example and we are pursuing his goals, trying to adopt the characteristics he embodied. Let me explore some of these characteristics because I feel they are essential to the success of a campaign such as ours or your own efforts to preserve treasured ecosystems. Listed in no particular order, I suggest the following: Commitment; Knowledge; Persistence; Thick Skin; Political Instincts; Organizational Skills; and Diplomacy.
>
> COMMITMENT: You either have it or you don't. George was appalled as a businessman and as a sportsman about how a public policy, such as the sugar program, was destroying such a vital resource. He was willing to commit his resources, energy and his time to his project. It became a focus of his being. His commitment was an educated and emotional commitment. He knew the uphill battle he faced. He knew the resources of the sugar industry and how hard sugar would fight to preserve the status quo. Those understandings were vital.
>
> KNOWLEDGE: He leaned on scientists to explain their findings to him. He became a para-hydrologist, a para–marine biologist, a para-agronomist. He became able to understand the reports

he was given so that he could take those reports to meetings and agencies and talk in the language of other scientists. He used that knowledge to take his case to federal and state policy makers. He used that knowledge to communicate with officials of various environmental groups, many of whom learned more about the seriousness of the problems from George. George educated himself about the sugar program and how it guaranteed profits to the sugar industry in Florida. Because he was so well versed on these issues, he made countless acquaintances at the EPA, the Department of Agriculture and the Corps of Engineers. His knowledge led him to be asked often to serve on boards, commissions, and advisory groups on Everglades issues, and he accepted many such offers. He, more often than not, became a force on those panels.

PERSISTENCE: This doesn't take much explaining. In short it means not taking "no" for an answer. Do not assume things are being done even though promises have been made. We must follow up on requests we make. Tasks and goals of bureaucracies get set aside, over-ruled, forgotten about, or studied to death. Do not hesitate to ask again, once you have been turned down. Persistence is critical.

THICK SKIN: Personally, I have learned how vital this is. I have been called an environmental extremist, anti-farmer and anti-job. I have been accused of leading an effort to force people off their lands. I have been portrayed as favoring alligators over family farmers. We become unfocused if we try to answer such name calling in kind. I have learned that whenever the name calling comes thick and fast we are making progress. We have our messages and our goals, and we must stay with them, no matter what we are called or how many setbacks we experience. We are not about hurting anyone, no matter what others may say.

POLITICAL INSTINCT: What George and I learned early on is that, even though we're Republicans, we needed to downplay party affiliation and instead label politicians as Pro Everglades or Potentially Pro Everglades. Try not to dismiss anyone. Try not to burn bridges. Today in our campaigns at both the federal and state levels, we have virtually an equal number of Democrat and Republican Representatives, Senators and state legislators who are our strong supporters. We have had various degrees of support from the Democrats in the White House and the Republican leadership in the Congress. We have received unexpected, and welcome, support from some quarters. We have countless advisors and team members of all parties. As I said, our labels are Pro Everglades and Potentially Pro Everglades. That is our politics. Nevertheless, it is essential to know how the Congress and the state legislatures function. We strive to know the key leaders in those bodies and so must you. In our cause we have benefitted from learning the politics of farm programs. We learned that there are reasons for farm groups to band together in an urban dominated Congress. That makes our job harder, but it is good to know what we face and what the limits are.

We need to know what avenues to use to get our message to Members of Congress or to the President. We are keenly aware that Florida is a vital state in the Presidential race. We would be crazy not to try to use that leverage. We have learned that educating constituents about our causes can impact their Congressmen or state legislators. Most politicians want to know what the voters are thinking. Polls sometimes get this message to elected officials, and we have done polling on our issue and announced the results.

While we may wish for campaign reform, we and the politicians live in an era where campaign contributions are a factor. Need I say more.

ORGANIZATIONAL SKILLS: Our group has depended on coalitions. Forming groups requires organization. Keeping your supporters requires organization. A good organizer must learn to delegate to others with organizational skills. Many are not naturally organized. It is important for those of us in this category to recognize that and to be sure we get someone who is organized to help us.

DIPLOMACY: We must treat even our opponents with respect. Our campaign has involved many debates, some on television. Being courteous and polite, while still getting our message across is so much better than going into a rage, pointing fingers, and making bitter accusations. It is good to see issues from other perspectives. Recognize that some agencies have their own problems. Do not demand more than some mid-level public official can deliver. Avoid burning bridges. One never knows when it may be necessary to go back to someone we may have previously alienated.

There are other traits, but I hope you get the gist of my message. Efforts to preserve environmental treasures are big undertakings. Our greatest rewards are preserving these wonderful areas for future generations. These future generations must realize that, with this gift of intact natural wonders, comes the same responsibility we have borne. Perhaps this is the most important message of all—the message to those who come after us.

HERE IS HOW THE penny-a-pound state constitutional amendments appeared on the November 1996 ballot:

AMENDMENT 4: Provides that the South Florida Water Management District shall levy an Everglades Sugar Fee of 1¢ per pound on raw sugar grown in the Everglades Agricultural Area to raise funds to be used, consistent with statutory law, for purposes of conservation and protection of natural resources and abatement of water pollution in the Everglades. The fee is imposed for twenty-five years.

AMENDMENT 5: The Constitution currently provides the authority for the abatement of water pollution. This proposal adds a provision to provide that those in the Everglades Agricultural Area who cause water pollution within the Everglades Protection Area or the Everglades Agricultural Area shall be primarily responsible for paying the costs of the abatement of that pollution.

AMENDMENT 6: Establishes an Everglades Trust Fund to be administered by the South Florida Water Management District for purposes of conservation and protection of natural resources and abatement of water pollution in the Everglades. The Everglades Trust Fund may be funded through any source, including gifts and state or federal funds.

A team of attorneys had toiled over the language. The attorneys began by taking up the 1994 state supreme court ruling that tossed George Barley's earlier amendment because, in the justices' opinion, it smacked of "political rhetoric" and involved more than one issue. Among the attorneys was Clay Henderson of the Florida Audubon Society, who described the process.

"It was agitated conversation because we're all passionate about the issue," he says, adding that there were "I would say countless emails, lots of back and forth, changes to the text, numerous conference calls and several face-to-face meetings, which is typical. A big issue was [the] 'polluter pay' [amendment] and whether or not it was self-executing, and that's a legal term, but all of the conversations since have been based on that issue. *Self-executing* means that it happens on the basis of the constitutional amendment by itself, and no additional legislation is required."

The attorneys toned down the language of the 1994 amendment and broke it into three measures to ensure each involved a single issue. The fee would have raised nearly $1 billion over twenty-five years. The campaign behind the effort, called Save Our Everglades, hired a manager and engaged a team of professionals and volunteers to gather signatures. Meanwhile growers pushed a rival amendment aimed at raising the number of votes required to approve any tax through a state constitutional amendment from 50 percent plus one to a two-thirds majority.

"We knew the Legislature would be unable and unwilling to engage in doing anything effective," recalls Charles Lee of the Florida Audubon Society. "We knew that the executive branch in Florida, the water management board, the governor, none of them were capable of escaping the grips of the control that the sugar industry is capable of exercising in Tallahassee. And so it came; it was relatively the same discussion that propelled the first effort on penny-a-pound, the idea that the only thing in the playbook when citizens cannot get relief through the Legislature, the only thing that remains other than court action, and again with court action you have to have basis to sue, the only game in town left is a constitutional amendment. . . .

"The belief was that the people of Florida would not stand for the inequity of everyone else having to pay to clean up their water quality problems, but the sugar industry was getting off scot free. And this was before—the [stormwater treatment areas] were in their infancy at this point in time—and we knew there was a lot more to be done. We didn't know how much there was to be done, and so having a major revenue source that would be funded from the sugar industry was seen as an important objective."

By August Save Our Everglades had collected enough signatures to prompt a state supreme court review of the amendment language. Representing the campaign was Jon Mills, Speaker of the state house of representatives from 1986 to 1988. Among the attorneys representing growers was Bruce Rogow, who had persuaded the justices in 1994 to find George's amendment unconstitutional, and Chesterfield Smith, president of the American Bar Association and chair in 1968 of the Constitutional Revision Commission, which had written the initiative petition law. Smith was known for having said about former president Richard Nixon, "No man is above the law."

The justices sided with Save Our Everglades only weeks before Election Day. "There is no confusion relative to who pays, how much they pay, how long they pay, to whom they pay and the general purpose of the payment," the unanimous decision read.

Mary had won where George had failed. Penny-a-pound was on the ballot.

"Today is a great day," she told a newspaper. "George is doing a jig up there, and he's probably having a drink, too."

AS ELECTION DAY NEARED, growers announced that the septic systems of Mary's and Paul's Islamorada homes were polluting Florida Bay with water that was dirtier than farm water, according to tests on samples they had collected, although a Save Our Everglades spokesman said the results were wrong. Meanwhile Save Our Everglades accused growers of sabotaging a $1,000-a-plate fund raiser at a Miami-area botanic garden that Robert F. Kennedy Jr. and other prominent environmentalists were scheduled to attend. Police ordered the event canceled after growers said they would bus one thousand demonstrators to the site for a $1-a-plate hotdog fund raiser.

Statewide, newspaper, radio, and TV ads and mailers and phone calls dogged Floridians. At one West Palm Beach TV station a sales manager said environmentalists and growers together had bought 140 advertising slots during the week before Election Day. That averages out to 40 ads a day, and with growers spending more than three times as much as environmentalists on radio and TV ads the debate shifted from the Everglades to government inefficiency and taxes. At one point Governor Lawton Chiles asked growers to pull an ad accusing the South Florida Water Management District and other agencies of wasting taxpayer money, but growers said no.

The barrage confused voters. Some believed the tax would be paid by Florida property owners, not growers. Others thought the tax would nudge up the price of sugar at grocery stores. Still others had their "yes" and "no" votes mixed up and thought voting "no" supported the tax. The confusion favored growers. Often when voters don't understand something they vote against it.

"We were increasingly despondent as we watched the updated polling numbers start to fall. We never saw a point before the election where it was going to go in the tank," remembers Charles Lee, "but the margin in support on Election Day if that deterioration continued was going to be very small at best."

Growers delayed their harvest so workers could door-knock in key South Florida areas. Growers especially focused on retirees, busing condo leaders from Broward, Dade, and Palm Beach Counties to the Everglades Agricultural

Area, where growers offered tours, hot lunches, and free sugar. Growers also focused on Black voters, convincing the Reverend Jesse Jackson to voice a radio ad urging minorities to oppose the tax. Some growers presented slide shows to civic organizations like the Kiwanis and Rotary clubs. Among them was Rick Roth of Roth Farms.

"I'm a certified single-engine private pilot, and I owned a plane at the time, and I actually did about ten or twelve speaking engagements," he says. "I remember specifically going to Rotary clubs and traveling an hour or an hour and a half to speak at civic organizations to tell our story. And I think the main story was, 'Look, everyone agrees we need to clean up the water that's going south because it's too much for the Everglades system. But the farmers have already committed to the Everglades Forever Act.' . . . We had a slide presentation, and I would go to these clubs and give a canned fifteen-minute speech and then have questions, and by the time we got done people were saying, 'Yeah, man. Give the farmers a chance. They're trying to do the job.' So I found myself being able to convince people because they did not know *why* the farmers were the bad guys. They just knew the farmers were the bad guys."

Roth explains the strategy.

"That was funny. The farmers wanted to argue, 'We're not the problem. We're not the ones polluting.' But the consultant said, 'You need to understand when you say to everyone, "I'm not polluting," all they hear is, "You're polluting."' So that's called in politics, 'If you're explaining, you're losing.' So if you're on the defensive trying to explain you're going to lose," he says.

"We need to be telling everyone this is not the way to solve environmental problems, by directly blaming agriculture for all of the ills. It's not just agriculture causing runoff of phosphorus. It's other people, too."

GROWERS ARGUED THEY COULD NOT afford a penny a pound, but could they?

Independent economists hired by federal and local water managers to study the issue concluded the fee would shave some 20 percent off grower profits of a nickel a pound but that the crop still would thrive in South Florida. The experts projected that some smaller farms might sell out to bigger companies but that sugar production would not be significantly affected.

"Personally it would have put me out of business. I actually bought 600 acres of farmland in January of 1996. I paid $3,500 an acre and bought farmland, and Roth Farms was already in debt but there was an opportunity and so I went for it," Roth says. "In ten years I would have been out of business. And maybe U.S. Sugar and Florida Crystals could have survived, but smaller farms like me would have gone out of business."

Roth's 4,000-acre farm at the time was among some 130 small and midsize farms in the Everglades Agricultural Area. His farm raised sod, vegetables, and some 2,000 acres of sugarcane and employed forty-five full-time employees. Roth believed he was already paying his fair share under the Everglades Forever Act. Penny a pound would have added to the burden. Roth was forty-three, married with three children, and president of the local farm bureau. The year before, his farm had failed to turn a profit as prices for leafy vegetables were way down.

"You were looking at this as it was going to put you out of business. It was very stressful. What helped me with it was I had been farming with my dad for ten years from 1976 to 1986 and then an additional ten years, so I had gone through a lot of booms and bust cycles," he says. "I'm just here to tell you that you just kind of build up a thick skin."

He says a big problem for growers was that they had just gotten started on cleanup under the Everglades Forever Act and lacked data showing the effort would work.

"That's what was so scary about 1996. We could not prove that we were the good guys," he says. "We had nothing."

MEANWHILE IN THE PRESIDENTIAL RACE, as Bill Clinton vied with Bob Dole for Florida's twenty-five electoral votes, Clinton signed a measure putting $124 million toward the Everglades. The Water Resources Development Act also included $10 million for Florida Bay, authorized the US Army Corps of Engineers to begin construction on the first of five stormwater treatment areas or filter marshes, and established a federal-state partnership to shoulder the cost.

"God created these places, but it is up to us to care for them," the president said during a White House ceremony. "Now we are, and we're doing it the right way by working together."

The following day Gore sent Mary Barley a letter. It read in part:

> President Clinton is committed to pursue a comprehensive plan to restore the Everglades and South Florida ecosystem. This rich, unique and fragile ecosystem is a symbol of what our nation's natural resources once were, and what they can be once again.
>
> That is why the President was pleased to sign important new legislation yesterday that will provide many of the tools we need for restoration and solidify our partnership with the State of Florida and its citizens in this noble effort. In addition, the Administration has worked with Congress to obtain funding and other resources required to restore the Everglades. The Administration also supports legislation that calls on Florida's sugarcane industry to contribute a penny per pound toward restoration costs.
>
> I salute you for your truly outstanding work on behalf of one of this Nation's greatest treasures. We realize that more needs to be done to help fund restoration. Therefore, I look forward to your continued involvement in this vital issue.
>
> Sincerely,
> Al Gore

RIGHT BEFORE ELECTION DAY, Mary Barley published the following op-ed in the *Naples Daily News*:

> On Nov. 5 citizens of Florida have the unique opportunity to vote to protect one of Florida's most precious and irreplaceable environments—The Everglades.
>
> At issue is who will pay to remedy problems associated with sugar cane production in the northern Everglades: Those sugar

growers who cause the problem, or Florida taxpayers? At stake is the economic health and vitality of an entire region, hundreds of millions of our tax dollars, and the future of an environment found nowhere else on earth.

The Everglades is part of the natural and cultural history of Florida. It is the source of clean, fresh water for over 5 million Floridians. Over 365,000 tourism-related jobs are dependent on the health of this ecosystem. The Everglades is the basis of the natural beauty which attracts millions of visitors to Florida each year.

For the last three decades Florida's sugar growers have dumped contaminated agricultural waste water into the Everglades, while diverting the vital fresh water that once nourished the system. Decades of abuse and irresponsibility on the part of Florida's sugar growers [have] resulted in an environment disaster that will cost between $1.6 [and] $2 billion to remedy. While these growers benefit from an extensive program of federal price supports and state water subsidies at a great cost to Florida taxpayers and consumers, they expect us to pay for a problem they created.

It's time we all said enough is enough! The Save our Everglades Committee is sponsoring three amendments to the Florida Constitution—Everglades Protection Amendments 4, 5 and 6.

Amendment 4 assesses a modest one-cent per pound fee on sugar produced in the Everglades; Amendment 5 prevents further pollution in the Everglades Agricultural Area; and Amendment 6 establishes the Everglades Trust Fund to ensure that funds go directly to Everglades restoration.

Vote "yes" on all of them.

ON ELECTION DAY THE Florida sugar harvest remained on hold as some twenty-five hundred workers turned out at polling places hoping to persuade any lingering undecided voters. Among them was farmer Rick Roth, who sat in a lawn chair in the rain holding up a sign that read "Vote No On 4."

"I'm a farmer. I'm prepared for the rain. I know what to do with the rain. Yellow is a farm worker jacket. We buy yellow rain jackets by the dozen," he says. "I actually was spit on. . . . It was a humiliating feeling being spit on."

That night dozens of Save Our Everglades supporters gathered in the moonlight on the rooftop of a downtown Miami office building overlooking the Miami River to await the results. The mood was festive as supporters sipped beer and wine, but the tax did not win.

Just over half of voters rejected Amendment 4 by a 10-point margin, although they approved the other two amendments, requiring polluters to pay for their own cleanup and establishing a trust fund. Together both sides had spent more than $35 million, with growers outspending environmentalists by more than two to one.

"Two out of three ain't bad," Mary proclaimed to supporters.

The results meant the Everglades Forever Act, loathed by environmentalists, would remain the primary source of funding for restoration. Meanwhile voters approved a rival amendment supported by growers making it harder to initiate a tax through a voter-supported state constitutional amendment, and they reelected Clinton. Here is how Mary remembers it.

"When we knew we were going to lose the penny-a-pound, we had to think what was our message and what we were going to say, and our message—and Charles Lee was the one who came up with it—we got two out of three. That's not bad. Yeah, we're disappointed, but we still got two, and the polluter pay is going to be an important constitutional amendment.

"I was very disappointed obviously," she adds. "It was just one of those things that we could have done better, and of course as soon as it was over we said, 'What we should have done was 'penny on the polluter,' not 'penny a pound.' But those are the things you learn, good things you learn. They could never have beat us had that been the message."

I wondered how George would have reacted.

"He would have been really upset, but then who knows? Maybe if he was alive we might have won," she says. "He would have thought that was a fair deal and that was what they should pay, and you know he would have been upset that the public didn't vote with him. It was the right thing to do."

12

THE FINANCIER

One reason the penny-a-pound political brawl of 1996 was so high-stakes was because of the vast amount of money invested by both sides in the issue.

Bob Buker of U.S. Sugar Corp. said the company was "borrowing money, as much as we can," as growers spent twice what environmentalists mustered. Mary Barley's campaign was funded almost entirely by a Wall Street tycoon who lived in Connecticut and resisted media attention. Many Floridians wondered who he was and why he cared so much for the Everglades.

In 1987 Paul Tudor Jones made more money than anyone else in America. He dumped stocks before a market crash, making between $80 million and $100 million. He was thirty-three.

He had arrived in New York after starting out in New Orleans when an uncle who traded cotton in Jones's hometown of Memphis, Tennessee, had gotten him a job with a friend. Jones then headed to the New York Cotton Exchange. By the time he was twenty-seven or twenty-eight he was making a seven-figure income, but he was bored. So he established Tudor Investment Corp., giving it the name because he felt it sounded better than Jones. Soon he was betting other people's money and getting great returns in the perilous world of commodity futures. His reputation grew as a wunderkind of Wall Street, and speculators flocked to his firm.

In 1988 the *Wall Street Journal* described Jones as the "most-watched, most-talked-about man on Wall Street" and sort of a playboy. He wooed clients with weekend hunting trips at his 3,272-acre wildlife preserve on Chesapeake Bay and threw elaborate parties at his estate, which included a formal garden, rare animal collection, and staff of three. He dated a young Australian model and speculated wildly on everything from cotton to foreign currencies, aided by a computer program nicknamed after Madonna. Jones kept a life-size cutout of the singer in his office. Sometimes while making million-dollar bets he wore Bruce Willis's old sneakers for luck.

In 1993, as George Barley was getting under way on penny-a-pound, a front-page article appeared in the *Palm Beach Post*. It began:

> Get used to the name Paul Tudor Jones II. Paul Tudor Jones II is one of Wall Street's shrewdest speculators and highest-paid traders. He is an avid fisherman and hunter with a $700,000 vacation home in the Florida Keys.
>
> And Paul Tudor Jones II is about to become a central figure in what promises to be a nasty and costly political battle over the imperiled Everglades. He will likely be depicted as both a benevolent hero whose generosity stretches from environmental causes to helping New York's needy and a greedy villain who will cost many Florida sugar workers their jobs.
>
> "I am hopeful he will be a major contributor," said George Barley, chairman of Save Our Everglades Inc., a political action committee seeking to convince voters that Florida sugar growers should pay a penny-a-pound tax for restoring the Everglades.

WHEN I REACHED PAUL TUDOR JONES in 2019 by phone, having arranged the call several weeks in advance through an assistant because of his busy schedule, the Wall Street tycoon traced his affection for the Everglades to his boyhood in Memphis. There he and his brother had spent most weekends outdoors with their father, himself an avid angler and hunter. A "love affair" with the outdoors evolved, Jones told me, and he began reading everything he could on fishing, hunting, hiking, exploring—Ernest Hemingway, Robert Ruark, anything on African safaris.

"I read every sporting magazine, outdoor magazine, any kind of magazine, anything that had to do with any outdoor adventure. I used to like to read all the stuff on Arctic exploration, just anything to do with the outdoors anywhere. I loved it," he said, with a drawl that stretched his words, twisted them as though they were rubber bands between two fingers. "If I could transport myself back, I'd transport myself back to the early 1800s. I'd be one of those explorers that would have been either in Africa or in the West. I would have loved to have been one of the first persons to Florida Bay. That would have been just magical. If I had a do-over in time it would probably be the early 1800s United States, because I think the United States is the most beautiful country in the whole world. And that would have been just heaven on Earth. If I could have been on a Lewis and Clark expedition, that would have been my highest dream of all time."

Paul Tudor Jones's fascination with Florida Bay began in 1968, when he traveled with his family to Islamorada during spring break.

> We drove down to the Keys. We stayed at the Cheeca Lodge. I'll never forget it. I hooked a barracuda on the edge of the dock. I was fourteen. I was really young. I remember bringing it in. It was a big barracuda, like 15 or 20 pounds. I remember bringing it all the way to the beach to unhook it because it was too heavy to pull out of the water, and all of these people were watching. I think it was one of the proudest moments of my life. I was so, so happy. And I'll never forget that. It's just an indelible childhood memory.

> For a little kid from Tennessee who had never really had a chance to be around the ocean that much, it was just for me, it was an enormous accomplishment. And I'd read all of these magazines, and so a barracuda might as well have been Jaws as far as a fourteen-year-old kid was concerned. So the idea that I got to hook that and fight it.
>
> And I remember it was a really long dock that had these tall pilings. And so I had to reach the rod around these pilings to pull the fish back toward shore. And so it was a very laborious process because that dock was probably 100 yards long. And when I finally got it there, it was probably after about 10 or 15 minutes, I kept thinking it was going to break off because I had a tiny little bass rod I'd brought from Memphis, a little Zebco rod.
>
> And so it was just a great feeling of accomplishment. And then I let the fish go, and it was just everybody was, you know, it was a happy ending to a great, great encounter.

IN 1986 AS AN adult Paul Tudor Jones watched a *60 Minutes* segment on Eugene Lang and his "I Have A Dream" program. Lang, a millionaire New York industrialist, had promised the graduating students of an East Harlem elementary school that if they stayed in school through college he would pay their college tuition. The rising Wall Street tycoon scheduled an appointment with Lang and later adopted a class of seventy-six students in Brooklyn's Bedford-Stuyvesant section, a poor, predominately Black area of New York City with a high dropout rate. More than half of the class went to college, and Jones adopted still more classes.

Jones invested $3 million in establishing the Robin Hood Foundation, a nonprofit that today still provides grants for some two hundred organizations dedicated to New York City's poor. Board members have included Katie Couric, the late John F. Kennedy Jr., *Rolling Stone* publisher Jann Wenner, and Children's Defense Fund president Marian Wright Edelman, a mentor to Hillary Clinton. In 1996 President Bill Clinton praised Jones and the foundation for its work on teen pregnancy.

Jones told me his dedication to the Everglades is inspired by a biblical sense of tithing.

"I go back to the Scripture I learned as a kid, because I think it's such a great guide book, though I would hasten to add I'm certainly not the best practitioner in the world," he said. "I would say part of the reason why the Everglades has been so important as a cause for me is, I just feel so blessed every time I'm outside. It brings me so much joy and happiness that trying to help preserve that ecosystem is one way of kind of squaring my debt with Mother Nature. Because she's obviously given ten thousand times more than I've been able to return the favor on. . . .

"The outdoors has been the root of all my happiness, if I'm being honest. It's been the root of all my happiness in everything I do with my family is generally outdoors, and we just have the best time ever. So I'm really grateful for any chance I get to help a good cause. It's going to hopefully pass this planet on to your generation and the next one in [as] a good shape as we found it."

IN THE LATE 1980S another vacation would change Paul Tudor Jones's life again. He traveled to England to hunt pheasants with friends and met a "tremendous prankster" named George Barley.

"George was about 10 years older than me, but I'd say emotionally he was about 20 years younger than me. And I'm probably emotionally backwards myself, so his childlike personality shown through immediately, which made me have a great affinity for him because he was mischievous and a prankster and funny and happy and just a little—the screws weren't completely tied on him. And that just was exactly my kind of person," Jones said. "At the same time, he obviously was a smart, really accomplished guy and a brilliant sportsman, great shot, great fisherman, and had a true and deep understanding of natural systems and an incredible bonding love for them. So that was like meeting my soulmate."

George Barley took Paul Jones tarpon fishing on Florida Bay, and they ended up way out near Sandy Key, the same spot where pea soup algae had sent George into a rage on his birthday. This time Jones says George was near tears.

"Why is this happening?" Jones asked.

"This is because of what the sugar industry is doing up near Lake Okeechobee, and it's depriving Florida Bay of much-needed freshwater," George explained. "And then what water is getting down here, there could be water quality problems with it, but I don't know. And I've got some guys, and there are some scientists, and they are working on it."

George was half right, of course. It was the Everglades' convoluted water management infrastructure that deprived Florida Bay of freshwater, although the infrastructure was convoluted because it was built to accommodate the Everglades Agricultural Area south of Lake Okeechobee, as well as Miami, Fort Lauderdale, and West Palm Beach, where breakneck growth was pressuring the fragile river of grass. Florida Bay was too far south to be affected by the cane fields' phosphorus pollution. Jones, a commodity trader at the time on the floor of the New York Cotton Exchange, right next to the Coffee, Sugar, and Cocoa Exchange, wanted to know more.

"Explain that to me," he said. "Explain to me the whole sugar thing."

When George Barley was finished, Paul Tudor Jones was angry, too.

"That's just wrong," he said.

"I know it is," Barley said.

"I'm going to help you on that," Jones said.

PAUL TUDOR JONES'S ENVIRONMENTAL RECORD was not perfect. In 1990 some 86 acres of destroyed wetlands at his Tudor Farms retreat in Maryland prompted a judgment of $2 million against him, the largest penalty ever at the time against an individual in an environmental criminal case.

Jones had bought the waterfront property in 1987. The property included 1,742 acres of wetlands and was situated northwest of the Black Water National Wildlife Refuge, a sanctuary for ducks and geese and home to one of the state's largest bald eagle populations. Jones wanted a duck-hunting preserve on his property and hired a project manager, who filled in the wetlands beside Chesapeake Bay and built roads, cleared woods, and dug ponds, all without any federal permits. When a US Army Corps of Engineers inspector found out about the work he ordered it all stopped until permits could be obtained. Instead the work continued, and the following year the inspector warned Jones he now faced civil and criminal penalties.

Jones fired the project manager, but by then the wetlands, once inhabited by endangered species and migratory birds, was what one federal prosecutor described as a "moonscape," with ponds for pen-raised ducks. In court Jones said he had not known about the missing permits and pleaded guilty to one misdemeanor count of violating the Clean Water Act. Prosecutors said he had been warned repeatedly to stop work at the site without the permits. Jones was ordered to pay a $1 million fine and another $1 million to the National Fish and Wildlife Foundation. He was also placed on probation for eighteen months, ordered to restore the wetlands, and in what the judge described as "restitution for the birds," barred from hunting wildfowl in the United States for a year. The project manager spent six months in prison. Jones hired someone else to finish the job and later became a director of the National Fish and Wildlife Foundation.

Tudor Investment Corp. also agreed in 1996 to pay $800,000 in fines to settle Securities and Exchange Commission (SEC) charges that the firm violated trading rules. It was the second largest penalty levied at the time by the SEC in a case that did not involve fraud allegations. The firm did not admit or deny wrongdoing in settling the civil case, which was filed in federal court.

WHEN GEORGE BARLEY DIED it was the absence of his energy that Paul Tudor Jones felt the most.

"When you lose a high-energy, funny, passionate person like that, on the spectrum of friends you might have, he was the 1 percent," Jones told me. "He was one of the brightest lights, if not the brightest light there was. And so that was like the flag bearer getting taken away."

On the phone with me the Wall Street tycoon recalled a Boyz II Men and Mariah Carey song, "One Sweet Day," that he said he listened to nearly every day on his way to and from work for a month after George's plane crashed. Here Jones grew quiet. I could hear him on the other end of the line searching on his computer for the song. Then I heard the song itself, the slow chords of the bitter-sweet pop ballad. Jones, in his lazy Tennessee drawl, began speaking the lyrics:

Never had I imagined living without your smile
Feeling and knowing you hear me it keeps me alive
And I know you're shining down on me from heaven
And I know eventually we'll be together
One sweet day

"It's just a great song," he said, mournfully.

BY 1996, AS Mary Barley's penny-a-pound campaign reached full intensity, Jones was forty-two and married (to the Australian model) with three girls. He wore round, wire-rim glasses, and his sand-colored hair was beginning to recede. He lived in an $11 million mansion in Greenwich, Connecticut, and maintained two retreats, including a $2.7 million home in Islamorada not far from Mary's. He was not unlike many in Florida who had come seeking endless sunshine and found instead that the paradise was less than it seemed. Jones's contributions on behalf of penny-a-pound were believed at the time to be the most anyone had ever given toward a political campaign in Florida history, but Audubon's Charles Lee believes his influence was even greater.

"More important than that was his contact list of people in New York and Washington and other financial centers around the country, who through his business and personal life he had detected were also either hunters or fishermen, or they had an emotive reason to be interested in protecting the environment," Lee says.

Jones infused the boards of the Everglades Foundation, the nonprofit organization he established with George, and other environmental groups with influential and wealthy business leaders and persuaded them to give, too. He also helped organize and unite the environmental groups by coordinating weekly conference calls, first called the "Paul calls" and later the "Barley calls."

"It was both his bringing his personal financial support and the financial support of his friends to the table," Lee says. "But it also was the skill he had to make the environmental organizations begin to think strategically and to begin to apply the same kind of business planning and campaign planning to their efforts that major business interests active in places like New York would bring to the table. He was a change agent that was successful in causing a much more professional and much more analytically driven approach to be adopted by the environmental groups that he influenced and came in contact with."

Mary Barley remembers Paul Tudor Jones as her greatest supporter after her husband's death.

"One hundred percent. Just everything. You have to remember he's putting millions of dollars in a campaign, and he's trusting me with his money to do the right thing. He never ever questioned me. To me that just says a lot about his trust, his values. It says that he knows me enough to believe that I would never take a penny from him, and he trusts me enough to tell him if I need help," she says. "It's not many people who would do that."

As Floridians were preparing to vote in October 1996 on penny-a-pound, the following op-ed, written by Paul Jones, appeared in the *Palm Beach Post*.

> Four years ago, the late Everglades activist George Barley said to me, "Did you know that we are actually paying the sugar industry to pollute Florida's Everglades?"
>
> George explained how problems associated with the harvest of sugar cane in the Everglades were killing this irreplaceable natural environment at a rate of 4 to 5 acres every day. George proposed that a 1-cent assessment on each pound of sugar grown in the Everglades be dedicated to repairing the damage done by decades of sugar's agricultural activity. The idea sounded fair and reasonable, so I looked into it.
>
> But first I had to answer this question: Was asking the sugar industry to pay justifiable? For the past 30 years, Florida's sugar industry has discharged agricultural waste water into the Everglades. Further, every scientist I spoke to and every report I could find detailed how the sugar industry is responsible for diverting much of the water and causing nearly all of the pollution in the northern Everglades.
>
> The cost of restoring areas damaged by the sugar industry could exceed $1.5 billion, according to an Oct. 9 report issued by the South Florida Water Management District. The sugar

industry admits to polluting the Everglades, continues to pollute it, yet it's paying only 15 percent of the cost of remedying problems it has created—$237 million, under the Everglades Forever Act, of the $1.5 billion needed.

Having been involved in the agricultural commodities business, I was in a unique position to evaluate whether sugar producers would be forced out of business by the tax. Since 1981, as a result of the federal sugar price-support program, sugar in the U.S. domestic market has sold for an average price of 23 cents per pound. Over this same period, the world price for sugar averaged about 11 cents. Our government actually sponsors a program that results in the American consumer paying twice the fair-market, wholesale price for sugar!

The sugar industry has been and still is among the most generous political donors in Florida, giving more than $10 million to elected officials. What does the industry get in return for its contributions? Elected officials, including Florida Sens. Bob Graham and Connie Mack, ensure continuation of the subsidy program.

According to the General Accounting Office, this subsidy costs everyone $6 per year. According to four different economists, this resulted in profits for 130 sugar-cane growers of about $4.9 billion since 1972. Other winners were Sens. Graham and Mack, who received more than $100,000 in contributions, and Rep. Mark Foley, who landed close to $50,000 this election cycle alone.

Who were the losers? Well, consumers received higher grocery bills, a degraded Everglades and a higher property-tax bill.

The sugar industry justifiably argues it should not be responsible for the entire cleanup. It says that the "billion-dollar tax," in combination with their Everglades Forever Act contributions of $237 million, will make them pay for a cleanup for which urban and industrial users also should pay. We agree. But remember, the sugar industry's payments under both the Everglades Forever Act and Amendment 4 will occur over a 25-year span. The present value of these payments actually amounts to 555 million 1996 dollars. That means even if Amendment 4 passes, the sugar industry will pay only 37 percent of the $1.5 billion cost.

On Tuesday, you will finally have the opportunity to vote on this issue. Save Our Everglades is sponsoring Everglades Protection Amendments 4, 5 and 6 to the Florida Constitution. Amendment 4 would require the sugar industry to pay a 1-cent assessment out of its federally protected 23-cent selling price for each pound of sugar produced in the Everglades. Amendment 5 would require Everglades polluters to pay 100 percent of the cost of cleaning their pollution. And Amendment 6 would start the Everglades Trust Fund to ensure that money meant for Everglades restoration will be spent on just that.

What will happen if Everglades Protection Amendments 4, 5 and 6 [pass]? Five million Floridians will have a source of clean, fresh drinking water. Property taxes will not be raised yet again to pay for the consequences of sugar industry agricultural practices. The problem of agricultural pollution in the northern Everglades will be resolved.

I encourage you to vote YES on Everglades Protection Amendments 4, 5 and 6. It is only a penny.

13

A BIG LAW

While George and Mary Barley and Paul Tudor Jones had faced off with sugar growers over penny-a-pound, water managers had toiled quietly over one of the largest man-made wetlands in the world, a 3,700-acre swath of cattails, water hyacinth, and water lettuce situated midway between Lake Okeechobee and West Palm Beach.

For two years the water managers had watched as the farm water, brown and murky, flowed into the marsh, the vegetative tissues absorbing the phosphorus runoff that was at the center of so much political debate. When the plants died their tissues released some of the phosphorus back to the water, but some remained fixed in the soil. Slowly the nutrient pollution in the water waned, and the water flowed brown and clear like tea as it once did.

The test marsh, authorized under the Everglades Forever Act—the legislation that had so angered environmentalists—was working better than expected, although the concentration was still near 20 parts per billion, higher than the target of a mere 10 parts per billion.

"No one knows how this is going to work completely," one water manager said.

At the time, the marsh represented the start of the largest ecological restoration ever, a $700 million effort. Within months bulldozers would begin clearing cane on some 66 square miles nearby, making way for six man-made marshes aimed at filtering the phosphorus out of the Everglades' water. Workers would plow gaps in the earthen dikes that held back the water, allowing the river of grass to once again flow in broad sheets. The workers would dig up muck 3 feet deep and blast through the limestone beneath while machines carved canals that would direct the water to the wetlands. The work was projected to take six years to complete.

The project was ambitious but addressed only part of the big picture. The man-made marshes were aimed at water quality, whereas previous restoration efforts had focused on the quantity flowing south into the parched Everglades National Park and Florida Bay. Meanwhile the US Army Corps of Engineers was engaged in a larger effort, the "restudy" of the Central and Southern Florida (C&SF) Project for Flood Control and Other Purposes, which had made way for modern Florida and brought the watershed to the brink. Congress in 1992 had called for the review as an update to the C&SF Project, designed for a population of 500,000. Today the population was projected to balloon to 3 million by 2010 and 10 million by 2050. (In fact the population has grown far more quickly than even those aggressive projections. By 2019 it was already nearly 9 million.) The engineers wanted to "replumb" the C&SF Project and resurrect a more natural flow in the river of grass, securing the region's drinking water for the future.

While water managers in December 1996 celebrated their initial success in the test marsh, taxpayers were shouldering most of the cost. Voters had rejected penny-a-pound, but that amendment had been one of three aimed at pressing sugar growers to pay more toward the effort. Voters had approved a related amendment requiring polluters to pay for their own cleanup.

Mary believed sugar growers were not paying their fair share.

IN 1997 FLORIDA'S SUPREME COURT JUSTICES wondered aloud how anyone could determine who was responsible for the Everglades' pollution and cleanup.

"We have Big Sugar, the ranches, golf courses and someone watering his lawn. All are polluters to some degree," Justice Leander Shaw observed. "How are you going to allocate who pays?"

Before the high court was Mary's "polluter pay" amendment. Governor Lawton Chiles had called for the hearing in Tallahassee to determine whether the amendment was "self-executing"—in other words, whether the state could implement the amendment without legislative action. The governor also had asked for a definition of the words "primarily responsible."

Months later the justices would rule that under the amendment polluters, and not taxpayers, were responsible for the Everglades' cleanup but that legislative action would be required to implement the amendment. The justices said that was because the amendment did not address how the state would identify polluters or define pollution.

Publicly Mary cheered the ruling. "My husband, George, would say this is a great day for the Everglades and a victory for the innocent taxpayers who are currently paying much more than their fair share," she said.

Privately the ruling represented another disappointment. She believed sugar growers had corrupted the state's highest court as they had its legislature.

"That was a real blow obviously to the campaign and to the hearts of all of us," she says.

"When you're in a campaign I'm not going to take on the justice system, because I know I'm going to be before them again, and in my mind I'm always thinking there are lots of ways I can get even along the way. Maybe I'll run into this guy, and I'll just tell him exactly what I think or something will come up and there will be a way to get even.

"And then there will be a time when he has to go to bed or she has to go to bed, and she'll be thinking about the things that she didn't do that she could have done or he could have done. And I heard that enough times from Nathaniel [Reed] to understand what that means, when you know you've done the wrong thing. . . . Because those are the things that when you go to bed at night you can't sleep.

"So, all that said, you have a public campaign and you want to keep the public with you. We're going to show, 'They think they've done us in. They've not.' . . . What you're telling them is, 'Don't give up because we're not ever going to give up.' And so your personal feelings are set aside for the good of the campaign."

The following spring the state legislature took no action on the amendment.

THE CAPE SABLE SEASIDE SPARROW is among the world's rarest birds, found only in Everglades National Park. That is, if you can find it. The drab-colored sparrow with a streak of yellow at the corner of each eye is no larger than the palm of your hand and shies from the flourish of flight. By 1998 this unassuming sparrow represented nearly all that was wrong with the Everglades.

The sparrow nests during the dry season a mere 6 inches from the ground. One nesting area was situated south of a series of floodgates and a levee that had shifted the river of grass's flow west, sparing farms and homes but flooding sparrow nests over and over. Other nesting areas to the east were so dry they burned. When heavy rain that year filled Lake Okeechobee, water managers protected the sparrow nests but in doing so were forced to inundate tree islands sacred to the Miccosukee tribe. Eventually water managers released the surge into the St. Lucie River, but that led to water quality problems that left fish with half of their flesh eaten off. Tourists fled.

Meanwhile the Cape Sable seaside sparrow continued to spiral toward extinction. One researcher estimated the population in 1992 at 6,400. By 1995 the population had plunged to 2,600.

"When they diverted the river of grass, they put it in the wrong place, and they made it flow at the wrong time of year," the researcher said.

In 1998 the US Army Corps of Engineers completed its C&SF restudy. A new restoration effort was emerging that was truly monumental. A half century after initiating the draining of the Everglades the federal government was considering an even larger effort aimed at doing the opposite—replenishing the watershed.

The scope of the situation was staggering. While the C&SF Project had been hailed in its day as the nation's largest public works project, this new, $7.8 billion plan was billed as the *world's* largest restoration effort. The plan called for much more water management infrastructure, and even then the river of grass would remain half of its former self, a fragment of the majestic 60-mile-wide swath that historically had swept south from Lake Okeechobee to Florida Bay. The plan would leave the C&SF Project largely intact but overlay it with a complex web of pumps, canals, dikes, wells, and pretty much every kind of reservoir the army corps could dream up. At the heart of the plan was water storage, rather than the drainage that wasted hundreds of billions of gallons at sea. The engineers said the plan represented the most ambitious attempt ever to restore a natural system. It spanned nearly 3,700 pages, enough to fill a four-foot stack of binders. "It is important to understand," the restudy explained,

> that the "restored" Everglades of the future will be different from any version of the Everglades that has existed in the past. While it certainly will be vastly superior to the current ecosystem, it will not completely match the pre-drainage system. This is not possible, in light of the irreversible physical changes that have [been] made to the ecosystem. It will be an Everglades that is smaller and somewhat differently arranged than the historic ecosystem. But it will be a successfully restored Everglades, because it will have recovered those hydrological and biological patterns which defined the original Everglades, and which made it unique among the world's wetland systems. It will become a place that kindles the wildness and richness of the former Everglades.

Vice President Al Gore, poised to run for president in 2000, unveiled the plan in Florida. "We've made restoration of the Everglades a national priority," he said.

The army corps claimed the plan would secure 85 percent of the 609 billion gallons drained annually into the sea. The plan consisted of a series of public works projects, each massive on its own and many considered risky because of the cost or reliance on water management techniques never tested on such a grand scale. The projects included one that would involve hundreds of deep wells, mostly around Lake Okeechobee, leading to a vast bubble of freshwater more than 1,000 feet underground. The freshwater would float upon the brackish water of the aquifer and, when needed, could be pumped to the surface. This project alone was projected to cost $1.5 billion. Never had so much water been stored underground, and some worried that only a small amount would be retrievable or that pressure from the pumping could fracture the subterranean rock.

All of the work was projected to take twenty years to complete, and the federal and state governments would share the cost evenly, although the plan contained no explanation of how the effort would be funded, including the projected $165 million cost of just keeping the Everglades running after all of the construction was done. Also, no one really knew whether the effort would work or how it would affect wildlife, such as the little Cape Sable seaside sparrow.

The plan was the work of an interdisciplinary team representing federal, state, tribal, and local government agencies, led by the US Army Corps of Engineers and South Florida Water Management District. The agencies said the work would position the region's drinking water for the future in the same way that draining the Everglades had made modern Florida possible.

The effort was aimed primarily at replenishing the Everglades through the removal of barriers and canals and the construction of reservoirs. Environmentalists and federal biologists worried that the effort did not go far enough toward restoring a more natural flow and that it prioritized agricultural and urban water users over the Everglades. They pressed the army corps to be more aggressive at dismantling the canals and dikes and restoring the river of grass's historic slow-moving sheet flow for the wildlife that depended on it. Utilities worried the untested water management techniques might drive up costs and jeopardize the drinking water supply if the techniques failed.

There was disagreement over what was natural. When the army corps considered removing part of a levee around the Arthur R. Marshall Loxahatchee National Wildlife Refuge, scientists there objected, reasoning that the refuge would dry out without the levee holding the water inside.

Perhaps the biggest barrier that had been inhibiting the river of grass's historic flow were the cane fields situated between Lake Okeechobee and the Everglades. For a decade sugar growers had resisted pressure by the federal government and of course the Barleys and Paul Tudor Jones to pay more for the cleanup. Yet again, perhaps no other group had more at stake than the growers responsible for a fifth of the nation's sugar.

The growers were concerned about the roughly 60,000 acres of farmland required to accommodate the reservoirs. The army corps indicated that the work would wipe out a seventh of the state's crop and 1,194 jobs over twenty years. The growers protested that under the Everglades Forever Act they had already halved the amount of phosphorus flowing from their fields with land management techniques called best management practices, which included tactics like dredging canals every year.

By the time the army corps redrew the plan to address all of these concerns it spanned more than 4,000 pages and ten volumes. Gore presented the plan to Congress in July 1999. Flanked in a meeting room steps from the Senate floor by a bipartisan group of Florida lawmakers, the vice president predicted future generations would look at the finished work with "awe and wonder." But first, Florida lawmakers would need to figure out where the state's funding would come from. That same month *Time* magazine honored Mary Barley as one of its "Heroes for the Planet."

"The Everglades is one of our most important natural cathedrals," she said. "It will be your legacy to the country."

TO BUILD CONSENSUS AFTER THE YEARS of bitter litigation and politics, Florida governor Jeb Bush appointed a commission consisting of some twenty members including environmentalists, municipal leaders, and sugar executives. Former state senator Curt Kiser, a friend of the Barleys, served as chair. I asked Kiser how everyone was able to get behind what soon would be called the Comprehensive Everglades Restoration Plan (CERP). "All you had to do," he told me,

> was look at what was going on down there, the quality of the water, the growth of the cattails taking over. . . .
>
> The sugar people, they were looking ahead and thinking, "If this whole thing becomes an environmental catastrophe who are

they going to blame? They're going to blame us, and then they may put us totally out of business." So it was in their best interest in the long run to help get behind the restoration project. . . .

I let everybody get all their gripes and bitches out. . . . I said, "Before we get started there's been a lot of history with everybody in this room, lawsuits, nasty articles, name-calling, all of that in the newspapers and on TV." So I let everybody get off their chests what it was that they were upset about. . . .

Likewise the municipal government people, the cities and counties, the water directors and county managers and city managers, they were as much a problem as anybody because they kept wanting assurances of how much more water they were going to get.

And I had to tell them, and Rock Salt was my Corps of Engineers guy, and finally I told them two or three times that this deal we're trying to cut here with CERP is not for you guys to get more water. Everglades National Park is starving for good clean water and we've got to do that. Florida Bay needs good clean water, and the environmental systems are the ones that this is for.

Now if in fixing their problems it makes more clean water available for drinking water, you know, fine. But that's not our purpose. That's not going to be one of the guiding principles that we work with. And when the different scientists and people got up and spoke about the problems and the history of it, how big it is and how much trouble it is going to be and how much time it's going to take and how much money it's going to take, after two or three meetings there was a pretty good consensus that everybody bought into. That, yeah, we need to do this, and the only way to pay for it is through a 50-50 deal. The state can't do it on their own, and the feds were not going to do it on their own.

It took several meetings for all this to kind of come together, but it did. And then we were able to put together a resolution that basically laid out the problems of the Everglades and said we needed a solution, and the only way to get this big restoration

> plan done was to have a partnership with the state and with the federal government. And that the resolution was a commitment by the commission, represented by all the different people on it, that we would join with the federal government and we would support the state of Florida paying 50 percent of the price tag. It passed unanimously.

IN JANUARY 2000 GOVERNOR BUSH CALLED on Congress to partner in the restoration as he announced the state's plan for paying the state's $4 billion share of the Everglades cleanup.

The governor said the state would spend $1.26 billion by 2010. An additional $1 billion would come from residents in Central and South Florida, but Bush declined to spell out how the money would be raised, leaving that decision to the South Florida Water Management District along with the region's municipal governments. The plan also did not explain where the money would come from after 2010, an estimated $1.7 billion. Bush said he wanted to avoid more taxes. Most environmentalists cheered the development, and so did sugar growers. But not Mary Barley.

"It's not a 10-year project. It's a 20-year project," she said. "Do you just say, 'I'm going to take out a 10-year mortgage . . . and 10 years from now I'm going to get the rest'?"

Mary went on to note that the governor also failed to "utter one syllable about enforcing the constitutional requirement that polluters pay their fair share of the cleanup."

The governor was a Miami developer who had not campaigned on environmental issues, and here he was poised to preside over the launch of the largest environmental restoration the world had yet seen. He responded that he was not aware the polluter pay amendment required legislative action but contended it did not apply here.

"Where that will come into play is on a separate corollary issue of great importance, and that is water quality," the governor said. "The restoration plan, by and large, relates to water capacity issues."

A few weeks later the following op-ed written by Mary Barley appeared in the *Tampa Tribune*:

The Tampa Tribune's editorial "Bush's solid Everglades plan" (Jan. 29) does not live up to the newspaper's reputation for taxpayer advocacy. In 1996, when the three Everglades protection amendments were before Florida voters, The Tampa Tribune wrote, "By voting for Amendments 4, 5 and 6, the people of Florida would forcefully act to save the Everglades and to force the major polluter to pay its fair share of the cleanup costs. The Tampa Tribune strongly supports the passage of Amendments, 4, 5 and 6" (Editorial, Oct. 18, 1996).

Voters followed the Tribune's advice and 68 percent, or more than 3.2 million Floridians, agreed that the "polluter pay" provision should be enshrined in our state's constitution. Once the amendment was approved, Gov. Lawton Chiles asked the Florida Supreme Court for an advisory opinion to determine what the term "primarily responsible" meant and whether the amendment was self-executing. The sugar industry tried to block Chiles' request for Supreme Court consideration ("Fanjuls Oppose State Court Hearing Pollution Case," Palm Beach Post, June 8, 1997). However, the governor prevailed and the Supreme Court issued its opinion in November 1997.

The high court held that in order for the polluter-pay provision to be given full force and effect, the state Legislature must enact enabling legislation. The court also held that the term "primarily responsible" meant that the polluters are 100 percent responsible for paying the costs associated with cleaning up their pollution.

Two sessions of the Florida Legislature have since come and gone, while the polluter-pay provision is still in limbo and the people's will continues to be ignored. A provision of the state's highest document, its constitution, remains but a toothless tiger because of the sugar industry's massive campaign contributions to Gov. Jeb Bush, members of the Cabinet, the state Legislature and both political parties.

The Everglades funding plan outlined by the governor last month makes it clear that he intends that South Florida taxpayers continue to subsidize the polluters. Restoration of the Florida Everglades will cost approximately $8 billion. The state's share of this cost is $4 billion over the next 20 years. Gov. Bush's proposal earmarks $1.25 million to be spent to fix the Everglades over the next 10 years. Where will the additional $750 million come from? The governor says from "South Florida resources." You don't have to be a rocket scientist to conclude that this is political doublespeak for tax increases.

The governor's spin is straight from the playbook of Otis Wragg, the dean of the polluters' spin doctors. At its best, this is hair splitting. Revenue flow is revenue flow, and if $1 billion flowed into the coffers of the South Florida Water Management District from the polluters, the need for other "South Florida resources" to fix the Everglades would be minimized.

We must be mindful that the Everglades supports the economy and quality of life enjoyed by more than 5 million South Floridians. It fuels an ever-expanding $16 billion–a-year tourism industry, providing 365,000 well-paying jobs. Most importantly, this jewel of nature is the source of fresh water for those of us who make South Florida our home.

It's time for the polluters to step up to the plate and pay their fair share of Everglades cleanup costs. It's high time for the stranglehold that this special interest wields over the politics of this state to end. And it is high time for the governor, members of the Florida Cabinet and the state Legislature to honor their oath "to support, protect and defend" the constitution of the state of Florida—or is placing one's hand on the Bible and taking an oath just another empty political gesture?

IT TOOK FIVE MINUTES, MAYBE TEN. President Bill Clinton used eighteen ceremonial pens.

A half century after Congress had authorized the draining of the Everglades, Clinton in December 2000 signed into law the congressional measure aimed at replenishing the watershed and saving one of the most threatened and treasured ecosystems in the world.

The $7.8 billion Comprehensive Everglades Restoration Plan that George Barley's friend Curt Kiser had helped craft was ambitious, but perhaps most remarkable was that it had brought together elected leaders, environmental groups, sugar growers, and tribal leaders who together had warred for more than a decade over the ailing watershed.

Gathered around Clinton in the Oval Office were more than a dozen elected leaders and environmentalists who had been most instrumental in pushing the plan through. Secretary of the Interior Bruce Babbitt and Senator Bob Graham were there. So was Governor Jeb Bush, but not his brother George W. Bush or Vice President Al Gore, who were monitoring the US Supreme Court hearing under way at the same time on Florida's hanging chads.

With the identity of his successor hanging in the balance Clinton grinned and leaned over his desk to sign the legislation. Standing over his left shoulder, peering down with an expression of patient satisfaction was Mary Barley.

"I was absolutely ecstatic because as you start working through all these things what you realize is, until the law changes nothing can change," Mary says. "The [US Army Corps of Engineers] has to do what it does because it is under federal law. And the [Environmental Protection Agency] and all the federal agencies had to live by the old Central and Southern Florida district laws, which favored agriculture 100 percent. And so CERP was a big, big step."

President Clinton signs the Comprehensive Everglades Restoration Plan of 2000 as Mary Barley watches over his shoulder

Courtesy of Mary Barley

Congress had approved the plan as part of the Water Resources Development Act, authorizing dozens of federal water projects across the country. The measure included a $1.4 billion down payment on fourteen of CERP's sixty-eight projects, which together were projected to be completed in thirty-six years. Here is some of what the plan included:

- 333 wells that together could store 1.7 billion gallons daily
- More than 200,000 acres of filter marshes, quarries, and reservoirs that together could hold as much as 540 billion gallons
- Removal of some 240 miles of canals and levees
- A 20-mile bridge across Tamiami Trail, US Highway 41's route through the Everglades, allowing the river of grass to flow beneath.

Construction was about to begin, but in fact the Everglades were far from saved.

14

THE POLITICS OF SCIENCE

How much was too much phosphorus in the Everglades?

Back in 1988, as US Attorney Dexter Lehtinen was preparing to sue the state of Florida over sugar growers' pollution in the Everglades he wondered, Exactly how much phosphorus could the river of grass withstand? He went to see Ron Jones, George Barley's friend. Jones reached up and pulled from a shelf a textbook on the science of freshwaters. The textbook contained a reference to a classic 1968 study that found that in watersheds like the Everglades, more than 5 or 10 parts per billion would begin to cause change. The matter was far from settled, however.

During the following years pretty much everyone would agree that the water flowing from the cane fields contained too much phosphorus. The amount measured as much as 300 parts per billion at its worst. In 1991 Governor Lawton Chiles announced to a federal courtroom, "The water is dirty! . . . I am here, and I brought my sword. I want to find out who I can give that sword to!" None of the court agreements or legislative actions specified, though, how much phosphorus actually was OK. That number—it would be called The Number—was left to scientists to figure out later.

Most scientists agreed The Number was between 0 and 50 parts per billion, a very small difference. Ten parts per billion represents 60 drops in an Olympic-sized swimming pool. At stake were millions of dollars in Everglades cleanup costs. Fifty parts per billion would be easier to achieve than five parts per billion. The debate would drag on many years after lawmakers approved the Comprehensive Everglades Restoration Plan, frustrating Mary Barley.

"Delay is the Everglades' enemy, and delay is the best thing that can happen to Sugar," she said at the time. "If you never get to a number, then you never have to clean it up."

RON JONES COULDN'T HELP BUT CHUCKLE in 2019 when I called and explained I was working on a book chapter about 10 parts per billion, and the chapter was titled "The Politics of Science." By then Jones had left Florida and taken up a research position in Oregon.

"It just became a matter of sanity over reality," he told me, "and I chose my sanity."

I wanted to know whether there were many other ecosystems in the world as sensitive to phosphorus as the Everglades.

"No, not that I'm aware of. There may be other ones, the Pantanal area in South America. There may be other wetlands that are oligotrophic like the Everglades, but none that I'm aware of," he said. "There have been so few studies done on any of this stuff. . . . People just don't work there, because there's nothing for them to see. They're measuring nothing practically."

The Everglades flourished in almost sterile conditions. Nourished by sunlight, water, and little else, the biodiversity thrived because of the paucity rather than in spite of it.

Biodiversity exists in an ecosystem because of conditions that prevent any single species from taking over. These conditions can include climate, disease, predators, or as in the river of grass, competition for sustenance. Each species finds an ecological niche in balance with all of the rest. The cattail is a native species of the Everglades, but a native species may become pernicious whenever a disturbance in the ecosystem upsets this balance. In the early 1990s phosphorus had caused cattails to spread across some 20,000 acres north of Everglades National Park, plus into smaller pockets inside the park and also the Arthur R. Marshall Loxahatchee National Wildlife Refuge.

After Lehtinen's lawsuit was settled state leaders agreed to reduce the flow of phosphorus, but the settlement did not establish any actual limit. A year later the settling parties agreed to plan for several studies aimed at identifying The Number. Ron Jones, Lehtinen's expert witness against the state, designed a study in which some of the watershed's most pristine water and sawgrass would be partitioned into narrow channels 100 meters long. He then would dose each channel with phosphorus and watch what happened, observing everything from large plants to microscopic enzymes. Sugar growers invested millions of dollars in another scientist, Curtis Richardson of the Wetland Center at Duke University, who undertook his own dosing study near the Everglades Agricultural Area south of Lake Okeechobee. Scientists funded by the federal agencies and South Florida Water Management District engaged in their own research.

Amid these efforts state lawmakers approved the Everglades Forever Act, which environmental groups blasted. What most intrigued Ron Jones about the measure was that until 2007 it allowed phosphorus of up to 50 parts per billion. This was not based on any environmental need—it was a political compromise.

Eventually the scientists funded by the National Park Service and US Fish and Wildlife Service concluded The Number was between 6 parts per billion and 10 parts per billion. Richardson, funded by the growers, said The Number was near 40 parts per billion, although he said the growers did not influence his research. Ron Jones was steadfast: The Number was 10 parts per billion. After serving as Lehtinen's expert witness against the state he had his funding withheld by the South Florida Water Management District—even though the state's own scientists funded by the same water management district came close to corroborating his stance. The state's scientists said The Number was between 10 and 15 parts per billion.

In court, attorneys representing the opposing parties worked to discredit Ron Jones. To testify against him they called to the stand other scientists, some of them friends of his who had introduced the midwesterner originally interested in marine biology to the Everglades. In government meetings officials publicly rebuked his work. After one South Florida Water Management District meeting, Jones went home and threw up.

Ron Jones persisted, he said at the time, out of a sense of moral obligation. He was a devout Apostolic Christian, part of a conservative movement with a doctrine that included a literal interpretation of the Bible. His marriage to his wife had been arranged after he had asked church elders for her hand. The couple never dated. Eventually he and his wife had six children.

"Work in the Everglades is something I feel is a calling," he said during a 2004 interview. "To me the Everglades is not just an interesting place. I believe there is a purpose for me being there."

BACK WHEN GEORGE BARLEY and Paul Tudor Jones began their advocacy work in the early 1990s, they scheduled a meeting with Donal O'Brien, chair of the National Audubon Society, and Bill Riley, an Audubon board member. Riley was a small-business investor who split his time between New York City, where his office was, and Florida, where he lived. He and his wife had published the *Guide to the National Wildlife Refuges*, which had been nominated for a Pulitzer Prize. Riley recalls meeting with George Barley and Paul Tudor Jones in Jones's New York City office. He says he told the businessmen-turned-advocates that he shared their concern for Florida Bay, but that real progress would be difficult without also addressing the Everglades' plight.

"Well, what would it take to do something about this?" Paul asked.

"Well, it would take, you know, considerable resources over considerable time," Riley replied.

"Well, I will commit three-quarters of a million dollars a year for five years," Paul said. "Will that get things started?"

"It certainly will," Riley replied.

Audubon used the money to establish a new position in Florida dedicated to the Everglades. George Barley, Riley, and Nathaniel Reed also contributed, but the advocates felt the position was not enough. The movement had the Everglades Coalition, an alliance of groups dedicated to the river of grass. But the advocates still believed the movement needed a leading organization focused full-time on the Everglades. In 1993 George Barley and Paul Tudor Jones established the Everglades Foundation as a research nonprofit. Riley was the founding director.

"We just felt that in the end the argument was going to be science," he says. "The focus was always on science, and obviously that science has to be translated into politics and appropriations and regulations and so forth, but unless it's science-based it's just—what's the point?"

The Barleys and Jones poured their fortunes into antisugar campaigns on behalf of Florida Bay and the Everglades because they believed in fighting money with money, but they also believed in science. They were convinced effective policy starts with scientific fact.

"If it's not science-based it's not going to be successful," Mary says. "That's the bottom line."

Mary Barley and Paul Tudor Jones at a 2019 Everglades Foundation benefit in Palm Beach
Courtesy of the Everglades Foundation

Today the Everglades Foundation disburses millions of dollars annually to groups like the Sierra Club and World Wildlife Fund. Its board of directors includes Jimmy Buffett, Jack Nicklaus, and Mary Barley. Mary also chairs the foundation's sister organization, the Everglades Trust, a nonprofit lobbying organization. The Barleys and Paul Tudor Jones believed that even if the Everglades' competing interests disagreed on politics, at least they should agree on what was proven as scientific fact.

"IT has been about the science from the very beginning!!!!" Paul Tudor Jones wrote me in an email. "There was so much fake news that we wanted to hire the best in the field to give honest answers and we ultimately did. From the very beginning we tried to find honest answers to all the environmental problems that beset the Everglades. And that is why so much of the state and federal policy has been driven by our scientists because they know we are 100% committed to scientific truth and nothing else."

BY THE EARLY 2000S the Everglades remained effectively engorged on phosphorus, with multiple restoration efforts having failed to prevent cattails from spreading across some 100,000 acres, pushing out the sawgrass the watershed was known for. Scientists estimated cattails were spreading at a rate of two to nine acres daily. The effort to rescue the Everglades from phosphorus and cattails effectively was one to rescue the river of grass itself.

Amid the overwhelming evidence, in 2001 the Florida Department of Environmental Protection sided with Ron Jones, along with environmental groups, federal scientists, and their own researchers, and announced The Number should be no more than 10 parts per billion. The new standard was poised to as much as double cleanup costs and extend the effort beyond a 2006 legal deadline. At the heart of the cleanup were six filter marshes, but it appeared they would not be enough. The phosphorus flowing from them measured at 20 or 30 parts per billion. Another option considered at the time involved enormous sludge-producing chemical treatment factories.

In addition to the quagmire around The Number, other issues were how or where the phosphorus would be measured, how pollution levels would be calculated, and what the state would do about violations. Nor was it known who would pay. Nearly two-thirds of voters in 1996 had approved Mary Barley's and Paul Tudor Jones's "polluter pays" amendment, but the legislature had never implemented it. Bills introduced in 1998 had died in committee. The measure remained little more than words on paper.

"Those in the Everglades Agricultural Area who cause water pollution within the Everglades Protection Area or the Everglades Agricultural Area," the amendment read, "shall be primarily responsible for paying the costs of the abatement of that pollution."

By 2002 environmental groups and the Miccosukee tribe, now represented by Dexter Lehtinen, charged in federal court that the state was badly falling behind on rescuing the Everglades' water quality. Lehtinen pointed to a study by Ron Jones, now a tribe consultant and director of the Florida International University Southeast Environmental Research Center, that showed pollution to be climbing on the western fringe of the Arthur R. Marshall Loxahatchee National Wildlife Refuge.

"What that means is that between 1989 and 1999 the Everglades have gotten worse, not better," Lehtinen said.

IN 2003 A BILLBOARD APPEARED on US Highway 1 in Islamorada featuring the smiling face of Mike Collins, a South Florida Water Management District board member, and his home phone number.

"Polluters love Mike," the billboard said. "Save the Bay, Save the Glades. Stop Mike Collins. Call MIKE COLLINS and tell him to STOP protecting polluters."

By mid-morning Collins, a fishing guide and thirty-year Islamorada resident, had unplugged his home phone. He characterized the billboard as "extortion."

"I'm not happy about it," he said. "My neighbors have seen it. They said, 'Hey Mike, why are you such a dirtbag?'"

Behind the billboard was Mary Barley's Everglades Trust. The billboard was over Collins's support of a measure moving through the legislature that would rewrite the 1994 Everglades Forever Act, which had infuriated environmental groups and spurred George Barley into action. The environmental groups charged this new measure was also bad, prompting President George W. Bush's administration to ask for changes when critics warned the measure might jeopardize his reelection in the crucial swing state if it looked as though the president and his brother Governor Jeb Bush had abandoned the Everglades. Mary warned that similar billboards might appear in other South Florida Water Management District board members' hometowns.

"Mike has turned his back on his hometown," she said. "He's promised to protect us. Our economy is totally dependent on tourism and the health of the Everglades."

A few weeks later Governor Jeb Bush quietly signed the measure into law, over objections from environmental groups, the Miccosukee Tribe, congressional leaders, and a federal judge overseeing the cleanup's progress, although the governor did ask legislators to change some of the newly signed measure's most troublesome language. The legislation maintained the 10-parts-per-billion standard but allowed the state to avoid a 2006 cleanup deadline, pushing back the deadline to as far as 2016. The deadline had already been delayed twice, but the governor said the new timeline reflected scientific reality, that pockets of the river of grass remained too polluted to be restored in three years. Monitoring stations would keep track of the water quality, but the data would be averaged in a way that some feared might hide pollution hot spots. Five years later a federal judge would strike down parts of the law, reasoning that the 2016 deadline was so far in the future it violated the Clean Water Act.

In 2013 Governor Rick Scott signed into law a new measure aimed at the Everglades' problematic water quality. The legislation put into action an $880 million plan the governor called "restoration strategies" that included another 6,500 acres of human-made filter marshes and two massive shallow basins to regulate flow. The work was scheduled to be completed by 2026.

THE WORK INVOLVED in getting the water down to 10 parts per billion involved two primary components. First, farmers had to stem the flow of phosphorus from their fields by 25 percent of what levels were between 1978 and 1988, before the cleanup began. To do this the farmers had to implement what were called best-management practices, such as using less fertilizer, changing irrigation techniques, and cleaning out drainage ditches. Variability in water quality was allowed based on rainfall. Second, the state had to get the phosphorus level down to 10 parts per billion using filter marshes. By 2016 the farmers had gone beyond 25 percent, achieving a 55 percent average reduction, and tests showed some 90 percent of the Everglades now met the 10-parts-per-billion phosphorus standard, the South Florida Water Management District said.

But the 25 percent reduction requirement was not applied to farms individually; it was applied regionally. This meant that if there were one or two farms where phosphorus levels were climbing, it would be OK as long as the region met the requirement. At one farm in 2016 there was a 210 percent average *increase* in phosphorus; at another there was a 141 percent average increase.

By 2018 the South Florida Water Management District was ready to declare the Everglades' water clean enough. In November the board voted to ask the federal court to vacate the 1992 consent decree and end federal oversight of the effort. Opposing the water management district in court filings the following January were the US Department of Justice, the Florida Department of Environmental Protection, the Miccosukee tribe, and environmental groups. Supporting the district were the Western Palm Beach County Farm Bureau, K. W. B. Farms, and Roth Farms.

The water management district argued that the consent decree hamstrung Everglades restoration. The district said it could get more water to Everglades National Park and Florida Bay but that the water would contain too much phosphorus, violating the consent decree and jeopardizing future permits for continued restoration. The district also considered the consent decree antiquated because the 1994 Everglades Forever Act went beyond the legal agreement's requirements. In February 2019 a federal judge ruled against the water management district. The consent decree and federal oversight of the Everglades' water quality remained in place.

IN 2018 GOVERNOR RICK SCOTT announced the state would commit $1 million toward an Everglades Foundation competition aimed at identifying new ways to cleanse water of phosphorus.

By now the Environmental Protection Agency described nutrient pollution as among the nation's most challenging and costly environmental problems. EPA scientists were observing a rise in phosphorus in lakes and streams across North America, even in remote, pristine environments. The most alarming impact took the form of harmful algae blooms. Not only had Scott declared states of emergency three times in the past five years over toxic blooms in Florida, but in Toledo, Ohio, a 2014 toxic bloom in Lake Erie disrupted the drinking water supply for some 400,000 people. In Oregon, harmful blooms in the Detroit River in 2018 prompted drinking water advisories affecting more than 150,000 people. In Utah, a 2016 toxic bloom in Utah Lake triggered hundreds of calls to Utah Poison Control. Some 130 callers reported physical symptoms after contact with the water such as vomiting, diarrhea, headaches, and rashes. There was even evidence of a widespread toxic bloom in the Alaskan Arctic, from the Bering Strait to the Chukchi and Beaufort Seas.

The Everglades Foundation said no affordable, scalable means existed for removing phosphorus from affected waterways. The foundation would award a $10 million innovation prize to anyone who succeeded in developing such a method. The organization said the multiyear international competition was modeled after incentive prizes that had encouraged Charles Lindbergh to make the first nonstop flight from New York to Paris and others that had led to the invention of fire extinguishers and commercial hydraulic turbines.

"If you couldn't buy the land, then you had to have another solution to make it work. You had to either build the largest sewer plant in the world to clean it up, and they weren't going to let you buy the land," Mary Barley says. "So this was a prize that could be the solution not just for us but worldwide."

The competition drew 104 contenders from thirteen countries and four continents, who eventually were winnowed to nine teams. One team from Pennsylvania proposed using nanotechnology. Another, from Toronto, had developed a mechanical-based, chemical-free system after processing hundreds of thousands of gallons of water daily for breweries. A team associated with the US Geological Survey aimed to recover the phosphorus and market it as a fertilizer.

The teams were tested on their ability to cleanse water of phosphorus down to 10 parts per billion. The Everglades Foundation named the competition the George Barley Water Prize. The donor of the $10 million was kept anonymous. The winner was to be announced in 2020.

15

RUNNING FOR OFFICE

In 2002 Mary Barley surprised everyone by switching her party affiliation from Republican to Democrat and jumping at the last minute into the race for Florida's commissioner of agriculture and consumer services.

She faced in the primary two little-known Democrats, a Miami public school teacher and librarian and an Orlando veterinarian. Running unopposed for the Republicans was incumbent Charles Bronson, a fifth-generation Florida rancher and millionaire who appeared poised for an easy victory. Mary's candidacy drew attention to an interesting fact about the cabinet position long controlled by agriculture interests: that the state's electorate was growing more urban.

"This is a department that's mandated to protect consumers, and they haven't considered the consumer in most of their decisions," the first-time candidate said. "It's time for a change."

When a newspaper journalist asked what George might think of her latest political battle she laughed hard. "He'd say, 'Go get 'em, Mary. And win,'" she said.

THE FLORIDA DEPARTMENT OF Agriculture and Consumer Services oversaw the state's $54 billion–a-year agriculture industry, the state's second largest after tourism. The department also oversaw thousands of acres leased from the state by sugar growers and agriculture growers' management of nutrient pollution. Consumer services ranged from car shops to travel agencies. Because of the position's convoluted name most Floridians call it simply "agriculture commissioner."

In 2002 the commissioner was poised to become more powerful after a recent change to the state constitution that reduced the cabinet's size from six positions to three: attorney general, chief financial officer, and agriculture commissioner. The cabinet shared executive branch responsibilities with the governor, and its reduced size would add weight to the agriculture commissioner's vote on statewide issues from environmental regulations to land development.

Mary ran on two main issues. She calculated that the department spent less than 10 percent of its annual budget on "straight up" consumer protection, although Bronson's campaign asserted the amount was closer to half of the budget because of food distribution and safety programs, pesticide regulations, and investigations into consumer complaints.

Mary also thought a new citrus canker eradication program was needed. Citrus canker is a disease that is harmless to humans but blemishes fruit and eventually weakens trees. An agriculture department program aimed at eradicating the disease, which at the time was spreading, had called not only for every infected tree to be cut down and burned but for every citrus tree within 1,900 feet of a sickened tree to be destroyed as well. South Florida landowners and local governments had mounted a successful legal challenge against the program, severely limiting it. Mary wanted to put a moratorium on the 1,900-foot rule, appoint a blue ribbon panel to study the disease, and establish a new plan.

Weeks after she launched her campaign Florida Citrus Mutual, an industry advocacy group, sent a letter to its 11,500 members describing her candidacy as an "alarming political threat." The letter announced that the organization's twenty-one directors had voted unanimously to tap an emergency fund and support Bronson. Funds "have been placed in reserve for dire emergencies when growers' survival is at risk," the letter continued. "This, unfortunately, is such an instance. . . . We can assure you that this candidate is a credible threat to your livelihood and should be taken seriously." The company cited concerns from citrus canker to nutrient pollution in waterways.

The letter was part of a growing campaign against Mary. A second group, calling itself Florida's Working Families, backed by citrus, sugar, and other agriculture interests, sponsored a TV ad that aired statewide blasting Mary as "barely a Democrat. Really a Republican." Shots flashed of Bob Dole and Newt Gingrich, whose campaigns Mary had never contributed to. The ad went on to say, "While Dole and Gingrich were cutting your Medicare, Mary Barley was bankrolling Republicans so they could give tax breaks to millionaires, like Mary Barley."

Meanwhile a direct mail flier featured images of computer printouts showing that Mary had contributed to other Republicans, including $1,000 for President George W. Bush's campaign in 1999. The flier said, "Millionaire Republican Developer Mary Barley. She's not one of us. She's one of them," although neither the ad nor the flier mentioned that Mary had also donated to Democrats, including $35,000 to the state party between 1996 and 1998.

"Who's ever seen an attack ad on an agriculture commissioner candidate?" asked one political scientist, who described the TV ad as a "hoot."

Florida's Working Families also paid for ads that aired in South Florida supporting rival David Nelson, the Miami public school teacher and librarian whose name was strikingly similar to that of US senator Bill Nelson, a well-known Democrat in the state. He also had the same name as the late son of TV icons Ozzie and Harriet. Mary was not surprised.

"I think it shows how important this race is to them, and I think it shows how corrupt the department is that these special interests are determined to keep it for themselves," she said. "The main difference is I am not a one-issue, special interest candidate. I will be representing Florida."

Mary's campaign for agriculture commissioner was short-lived. She lost in the primary to Nelson, drawing 35 percent of the vote to Nelson's 44 percent. Going into the general election Bronson had raised $1.46 million, while Nelson had $11,279.

"It was a fun campaign," Mary says. "I love having to make Sugar spend money. I've cost them millions and millions of dollars, and I love every dollar that I've cost them because if there is anybody who is greedy and disgusting, they are."

MORE THAN ONCE during the ten years or so I spent working on this book, as I listened to Mary Barley describe her all-consuming campaigns, I asked whether she enjoyed politics. Invariably her answer was the same—not really.

"I wouldn't say I enjoy it. It's something I'm passionate about, that I believe in, that I think needs to be done," she told me. "Somebody had to do it, and I'm willing to do it if nobody else is going to do it. . . . The other side of the coin is I would not give it up."

Motivating her instead, as it did her late husband, was an utterly intractable moral determination to hold accountable those she judged as corrupt. There was no gray area here; only black and white. Good was good, and bad was bad. That was that, and on her spectrum sugar growers were evil.

"When I was on the campaign trail they were just as mean and evil," she said. "They would get right into my face and right into my space. They're all trained in that. There is a space where you feel threatened, and they get right into that space."

Mary Barley was a woman who was sure of herself and sure of her truth.

"I know it's hard for you to understand, but I'm not intimidated by this," she said. "I mean, I know where I come from, but maybe that's the reason I feel so like, You're just another guy, and if you're not producing and acting like you should in the Oval Office I'm not going to be that impressed with you."

I admired her for this quality, even if I did not always agree with her. Perhaps our biggest point of disagreement was my journalistic responsibility to talk to the growers, to listen to their side of the story. If it was truth I was after, she said, I would not find it there, and we argued. I was letting the growers off easy, she said. I was not telling the whole story.

"To me that is not the story," she said. "The story is how evil and greedy and selfish they were and are, and if they could have been good corporate citizens this would all be done, and we would be well on our way to—probably restoration would be done or darn close to it. So every time I see something nice about them I have to grit my teeth.

"It's not about being biased. It's just telling the facts. I'm not biased. I can't say anything good about Sugar. I can't think of one thing they have done that they haven't been forced to do by the court of law, mainly, or by legislation. The legislature wouldn't have done anything without the court of law, so nobody was out to do the right thing for the people. And so to me that is not being biased. That is just a matter of fact, and that's how it was....

"They don't want to deal. They're not going to give anything up. You do what they want, and they're not going to give an inch."

WITH MARY OUT OF THE RACE for agriculture commissioner, the contrast between candidates leading up to the November 2002 elections could not have been starker.

Republican incumbent Charles Bronson was a veteran of state politics. He had lost two previous races for agriculture commissioner before he was elected in 1994 to the state senate, where he served until 2001, when Governor Jeb Bush finally appointed him to the position he had long sought. The previous agriculture commissioner had stepped down to take another position.

Having raised more than $1 million for his 2002 campaign, Bronson planned an extensive advertising blitz. The TV ads would highlight his commitment to consumer issues like price gouging and portray his opponent, David Nelson, as an unqualified school librarian.

Nelson was in fact engaged in his first bid for public office. He described himself as "Joe Novice" as he raised less than $20,000. He folded his own brochures, and after the school year began in August he reserved his campaigning for nights and weekends because, he said, "my mortgage company doesn't understand when I'm not working." Polls showed he either was leading or running neck-and-neck with Bronson, and many believed it was because of his name.

Agriculture interests that had supported David Nelson against Mary now got behind Bronson, including Florida Citrus Mutual. Despite their ads ridiculing Mary for being "barely a Democrat," party affiliation apparently was not all that important to them after all. In the end Bronson won the race, making way for the Republicans' first clean sweep of the state's newly downsized cabinet.

"It was a clean race that focused on the issues, and I feel very good about that," Bronson said.

David Nelson said he would return to his job at his Miami middle school, where he would advise the chess club and National Junior Honor Society.

"I think I did pretty good for $15,000," he said.

As for Mary, she felt relieved.

"Relieved when I had my life back again," she says.

16

A BIG DEAL

"I have an idea that might solve all our problems," Governor Charlie Crist said in 2007 during a meeting with two U.S. Sugar Corp. lobbyists. "Why don't we just buy you out?"

The governor, a Republican who had been in office for less than a year, continued.

"If Sugar is polluting the Everglades, and we're paying to clean the Everglades, why don't we just get rid of Sugar?" he asked.

In June 2008 jaws dropped as Crist announced the state had agreed to a $1.75 billion deal to buy out U.S. Sugar, the nation's oldest and largest sugar producer, and put the land toward Everglades restoration. The agreement included the company's 187,000 acres, railroad, citrus plant, and $100 million state-of-the-art mill, which had opened four years earlier. The land acquisition would have been the largest in state history. U.S. Sugar would go out of business in six years. Crist stood at the edge of the Arthur R. Marshall Loxahatchee National Wildlife Refuge, the Florida sun beating down, as he declared the acquisition as historically "monumental as the creation of the nation's first national park, Yellowstone."

"I can envision no better gift to the Everglades, the people of Florida, and the people of America—as well as our planet—than to place in public ownership this missing link that represents the key to true restoration," he said.

In a strange moment Robert Coker, senior vice president of U.S. Sugar, and Mary Barley shook hands. Mary recalls the moment now with incredulity.

"I'm thinking, 'Are you kidding me?'" she says. "They didn't deserve to be applauded because they weren't doing it for the good of the state. They were doing it for themselves, as always, and to me it was just another way for Sugar to make people think they were doing the right thing when they're not. . . . They're evil people. That's all I can say about them. They don't care about anything except the almighty dollar."

THE "MISSING LINK" Crist spoke of was the river of grass's long-lost flow from hydrologic heart Lake Okeechobee to Everglades National Park and Florida Bay. The Comprehensive Everglades Restoration Plan (CERP) had glaringly left out any such link, relying instead on massive public works projects that would move water around and even underneath the untouchable Everglades Agricultural Area. Finally, the governor appeared to be implying, this historic flow would be restored, and Florida's treasured river of grass would be resurrected from its vast green coffin of cane fields, realizing a dream long held by environmental groups, which hailed the deal as historic.

"A restored and sustained Everglades is no longer a dream," Mary Barley said.

Under the deal U.S. Sugar would sell its land to the South Florida Water Management District, which then would swap tracts with other growers to assemble a large swath. Although Crist was touting a vision of a renewed river of grass, the swath would not be used to revive the watershed's historic flow, not precisely. By now the land was too saturated with fertilizers and sunken from erosion. Instead the swath would be used to address the restoration effort's two largest challenges: there wasn't enough freshwater, and the water was too polluted.

State leaders said this likely would mean big changes for CERP as new projects like filter marshes and reservoirs would make existing projects unnecessary. The environmental groups especially hoped a plan for some three hundred wells leading to an underground reservoir, an unproven concept, would be scrapped. State leaders planned to sell U.S. Sugar's other assets to remaining growers.

News of the deal came as a shock to almost everyone except top leaders of the Florida Department of Environmental Protection, the South Florida Water Management District, and U.S. Sugar and Mary Barley, Paul Tudor Jones, and the Everglades Foundation, which had been involved in the secret negotiations from the beginning. At Crist's June announcement foundation representatives were ready with gleaming press kits touting the proposal.

"We've been asking governors to buy them off for a long time," Mary said.

Weeks before Crist's 2006 election Paul Tudor Jones—his net worth then valued near $3.3 billion, making him among the richest in America—contributed $405,000 to the state Republican Party, the largest single donation the party ever had received. A few months later the newly installed governor joined Jones on a fishing trip on Florida Bay. While casting for snook the Wall Street tycoon urged Crist to revamp the board of the South Florida Water Management District, the state agency overseeing Everglades restoration, by including members favored by environmental groups rather than sugar growers.

Two months later Crist made two key appointments to the board, including Shannon Estenoz, a civil engineer and former Everglades Foundation board member. When growers, fearing that a low lake level might imperil their crops' water supply, asked the water management district for permission to pump their polluted water back into Lake Okeechobee, the board said, perhaps unexpectedly, no—not when the state was spending so much to clean up the Everglades.

That prompted U.S. Sugar's Robert Coker to send two lobbyists to meet with Crist—Brian Ballard and Mac Stipanovich. Coker told a newspaper he was "stunned" by the governor's buyout suggestion and took it back to the company's board, which did not reject it.

"When you own something and build something for 80 years," Coker said, "you develop an emotional attachment to the business and to the land. The descendants of Charles Stewart Mott, who make up the majority of our board, have had offers in the past for all or parts of our company and our land. They never felt it met their criteria.

"We believe that our company and our board and our shareholders have gotten two things. We've gotten reasonable fair value, not what we thought we could have gotten. . . . And at the same time they know these lands are going to be used to ensure the future of the Florida Everglades. I think that's a legacy they were comfortable with."

In February 2008 Crist asked his staff to look into acquiring U.S. Sugar's land. The Everglades Foundation's Tom Van Lent provided models showing how the land could be used most effectively. In fact the foundation was a "fantastic source of not only support but information," Michael Sole, secretary of the state Department of Environmental Protection, told me at the time.

PERHAPS MOST SHOCKED by the news were the residents of the Everglades Agricultural Area, especially in Clewiston, where U.S. Sugar provided 25 percent of the tax base and was responsible for more than half of the economy.

"The people in Clewiston were the ones who were most upset because U.S. Sugar had always had kind of a paternalistic view of Clewiston," farmer Rick Roth says. "That was the hometown of U.S. Sugar, and U.S. Sugar is a huge corporation with a couple hundred thousand acres of farmland. And the baseball teams, and the Little Leagues and the Rotary club, and everyone depends on corporate contributions. . . . So I think that was the big shock, was the regular folks who live in Clewiston. This just couldn't be happening, because that just means Clewiston was going to dry up and blow away."

U.S. Sugar had been responsible for the community's auditorium, library, pool, and youth center and paid for the college educations of many of its employees' children. Some employees had worked there for many decades and considered U.S. Sugar like family. Under the buyout plan in Hendry County, where Clewiston was located, some 30 percent of the land would be owned by the state or federal government, jeopardizing 24 percent of the county's tax revenue. Homeowners feared property values would plunge, trapping them where there were few jobs. Soon after the announcement the First Baptist Church of Clewiston, on a sign out front, announced the theme of the Sunday sermon: "A word for the anxious, angry and fearful."

Clewiston and Hendry County leaders arranged an emergency meeting, where they voted to hire an attorney and commission an economic impact study. They also called on the state to provide an economic transition and support plan. A University of Florida study concluded sugar growing would continue in the Everglades Agricultural Area, as other growers acquired U.S. Sugar's assets and maintained production. The report also found the domestic market would not be significantly affected, because of the federal sugar price-support program. The report noted the program kept the price well above world prices, although prices appeared to be converging over time, but this was little consolation for many in the region.

"This is pretty normal, the last thirty-five years fighting for our right to farm in the Everglades," Roth says. "It's just another fight. It's just a different kind of fight. You just keep coming up with a different way to try to put us out of business, and this is just another way of trying to do it."

WEEKS BEFORE GOVERNOR CHARLIE CRIST'S staggering announcement construction had been under way not far from the Arthur R. Marshall Loxahatchee National Wildlife Refuge on what would have been the largest reservoir of its kind in the world.

Bulldozers and dump trucks rumbled through the swamp as they worked to remove some 30 million tons of dirt and muck for the $800 million reservoir, the largest and most expensive part of CERP. The Manhattan-sized reservoir would have been so large, someone standing at one end would not have been able to see across to the other, and officials were considering whether to allow boating and fishing after the projected completion in 2010.

The A1 reservoir, as it was called, was being constructed on land acquired under the Talisman Sugar Corp. buyout. The reservoir would have held up to 62 billion gallons of water, enough to fill 100,000 Olympic-sized swimming pools, helping to restore the river of grass's historic flow south and alleviating the need for harmful releases east and west to the St. Lucie and Caloosahatchee Rivers, where the water threatened delicate estuaries. What made the reservoir unique was its above-ground construction. Most are built within canyons or valleys and rely on natural water sources like a river filling behind a dam. This reservoir would stand on its own, fortified with earthen-and-concrete walls some 26 feet high. Eventually the plan was to double its size.

The reservoir was among more than a dozen CERP projects that were now suspended or canceled as attention now turned to the U.S. Sugar buyout. Some $1.3 billion had been spent on the projects.

From the beginning some voiced concern the deal diverted funds from existing Everglades projects. Soon after Crist's announcement the Miccosukee Tribe filed a lawsuit arguing the purchase would delay restoration. Eventually the state would pay a contractor hundreds of millions of dollars over the stopped reservoir work. Others argued the scattered parcels hardly constituted a "missing link" and questioned the timing. In 2008 Crist was considered a potential vice-presidential running mate with US senator John McCain, who would lose the election to Barack Obama.

FLORIDA'S HISTORIC BUYOUT of U.S. Sugar for Everglades restoration would not happen. The deal had been struck on the cusp of the worst recession in a generation. Across Florida homes were falling into foreclosure as the real estate market collapsed. Five months after Governor Crist touted a land acquisition as monumental as Yellowstone National Park, the agreement was downsized to $1.34 billion for 180,000 acres, excluding U.S. Sugar's assets. Crist described the new deal as "miraculous." By April 2009, with the recession deepening, the deal was downsized again to $536 million for 72,800 acres. U.S. Sugar would stay in business, and the state would retain options for purchasing the remaining 107,000 acres in the next ten years. Even this, much more modest acquisition would have been the largest ever for the Everglades.

By 2010 the deal appeared poised to some to rescue U.S. Sugar more than the river of grass. Competitive pressures and operational problems had pushed the company's debt in 2007 to more than $500 million. The company was also suffering through an extended drought and faced a shareholder lawsuit. Chief competitor Florida Crystals, owned by the Fanjul brothers, characterized the deal as a bailout, a strong word after the federal government had come to the aid of the same financial institutions that had caused the economic collapse. One *New York Times* article went as far as to suggest U.S. Sugar was dictating a large measure of the negotiations.

The corporate law firm representing the company was Gunster, whose chairman was George LeMieux, Crist's chief of staff at the time the buyout deal was conceived. LeMieux returned to Gunster in January 2008. LeMieux and Crist were confidants; the governor called LeMieux the "maestro" of his 2006 election victory, and in 2009 when a US Senate seat was vacated Crist appointed LeMieux to the office. At Gunster LeMieux said he recused himself from the U.S. Sugar negotiations to avoid any appearance of a conflict of interest.

Under the latest deal the South Florida Water Management District would be left with six large disconnected parcels, including all of U.S. Sugar's citrus groves. State leaders defended the negotiations, saying U.S. Sugar would need certain tracts to continue farming. Meanwhile internal documents obtained by the *Times* suggested the price the state was offering U.S. Sugar for its land was inflated and based on values from 2004 to 2008, when prices had been much higher, rather than 2009, when prices were far lower. Restoration costs, projected to reach into the billions of dollars, would be on top of that, and the documents suggested the water management district had nowhere near that much money.

In October 2010 the water management district closed on 26,800 acres for $197 million. The acquisition represented a seventh of what Crist had proposed and included 17,900 acres of citrus groves and 8,900 acres of cane fields. The deal preserved options for future purchases, including

- An exclusive three-year option to purchase either a specified 46,800 acres or the entire 153,200 acres at a fixed price of $7,400 an acre,
- A subsequent nonexclusive two-year option to purchase the 46,800 at fair market value, and
- A subsequent nonexclusive seven-year option to purchase the remaining acres at fair market value.

"I'm surprised I didn't cry, really," Mary says.

The water management district cited a $150 million decline in revenue since 2008, and district leaders said this was all the state could afford. Kirk Fordham, chief executive officer of the Everglades Foundation, suggested the state pursue other funding sources, namely sugar growers.

"The burden continues to fall on the taxpayer," he said. "It's our hope that through this process there will be a shift in attention to those that have been creating the problem."

17

TODAY'S BIG PICTURE

Few species are more iconic of the Everglades than the American alligator. Certainly the Burmese python is a rival, but it is not a species of the Everglades. Instead the python is an interloper here, a freeloader, the toxic serpent in the Garden of Eden, although rather than dust, it consumes almost every living creature it encounters in this long-suffering paradise.

The Burmese python is perhaps the most infamous invasive species of the Everglades. The pythons are native to Southeast Asia and were imported to the United States as pets. Many of the tens of thousands slithering among the marshes were released by owners who, perhaps after falling for their beautiful skin patterned with diamonds of deep mocha and rich gold, grew fatigued by their enormous size. The pythons are among the largest snakes on Earth, measuring 23 feet or more in length and weighing up to 200 pounds, with a girth as large as a telephone pole. The carnivores are not venomous but squeeze their victims until the prey suffocate. Stretchy ligaments in their jaws enable the pythons to swallow their victims whole.

It is no coincidence that as Burmese pythons in the Everglades have flourished in recent decades—estimates of their population run into the hundreds of thousands—populations of other animals have declined radically. Scientists suspect pythons are responsible for the disappearance of up to 99 percent of rabbits, raccoons, and other small mammals in Everglades National Park. The South Florida Water Management District announced in 2019 it would triple its budget for its python elimination program to $1 billion and double its number of contracted hunters to fifty. The district will not have any trouble filling the jobs; some twenty-six hundred hopefuls applied from across the globe.

The American alligator, on the other hand, truly is right at home in the Everglades. With armored, lizard-like bodies, muscular tails, and powerful jaws the alligators look as though they are creatures of a distant past, and they are. The species is more than 150 million years old, having survived the mass extinction some 65 million years ago of their original contemporaries, the dinosaurs. Perhaps this remarkable resilience is the reason for their smirking expressions, a sense of superiority born of long experience. Perhaps it is because they know something you may not: the American alligator is not the only crocodilian calling the Everglades home.

The Everglades represent the only place on Earth where freshwater alligators coexist with saltwater crocodiles. This makes the watershed the only place on Earth where one really can say, "See ya later, alligator. After a while, crocodile," as one biologist told me. Perhaps the alligators and crocodiles are all smirking because they are in on the joke. In any case the American crocodile is most easily distinguished from its alligator cousin by its longer, thinner snout; its lighter, olive-green color, compared with the alligator's black color; and the two long teeth on the lower jaw that are visible when the mouth is closed.

Alligators and, perhaps to a lesser extent, crocodiles are emblems of the Everglades in part because scientists have designated them so. In the Everglades crocodilians are ecological indicators, monitored as a measure of the watershed's well-being and as a tool for communicating the scientific findings to decision makers and the public. Scientists concentrate on ecological indicators because it would be impossible for them to monitor every animal and plant in the Everglades all the time. Indicator species are selected based on the breadth of their impact on the ecosystem.

Alligators, for instance, are responsive to environmental changes and are influential as top predators and ecosystem engineers, forging holes, trails, and nests that provide habitat for other animals and plants. The alligators are easy for decision makers and the public to understand and identify with. How are the Everglades doing? Is restoration working? The smirking alligators and other ecological indicators, like invasive exotic plants (the melaleuca tree and Old World climbing fern) and elegant wading birds (the white ibis and wood stork), will tell us.

The answer is that alligators in the Everglades are not doing great. They weigh 80 percent of what they should. They grow more slowly, reproduce less, and die younger than other alligators. What the alligators and other ecological indicators are telling us is that some twenty years into the Comprehensive Everglades Restoration Plan, much work remains to be done.

IN THE 1960S A DEVELOPMENT COMPANY purchased 57,000 acres of swamp in the Picayune Strand southeast of Naples and went about building the world's largest residential subdivision.

Strands are geologic features of the Everglades formed by linear bedrock depressions. Soaring cypress trees distinguished these swamp corridors before logging. Today they are characterized by hardwood swamp trees. Botanists value them for their unusually rich subtropical foliage, most notably orchids, bromeliads, peperomias, and ferns.

Gulf American Land Corp. constructed canals partially draining the land, laid down miles of roads, and sold lots to long-distance buyers who did not know that the land would flood during the rainy season. Eventually the development failed and the company went bankrupt, leaving behind the roads and canals that overdrained the area, reducing its ability to recharge its aquifer, a drinking water source.

Undoing the remains of the Picayune Strand development was the first CERP project to begin construction in 2007. The project encompasses nearly 55,000 acres and involves removing some 260 miles of crumbling roads and filling in 48 miles of canals. Pump stations will enable the water to flow again, replenishing the aquifer and adjacent Fakahatchee Strand State Preserve. The project will also restore habitat for vulnerable species, including the endangered Florida panther, the official state animal with a population of between 120 and 230 left in the wild. The project is scheduled to be complete by 2023.

During the very first years after CERP became law, not a single project was constructed as costs rose. By 2004 the effort was $1 billion over budget and two years behind schedule, and the unprecedented federal-state partnership was faltering. Leaders had agreed to share costs evenly, but with Congress failing to approve any funding Governor Jeb Bush was left to put $1.5 billion of the state's own money toward a program called Acceler8, aimed at jump-starting eight projects including the Picayune Strand and a reservoir in the Everglades Agricultural Area.

"We have missed almost every milestone," CERP's project manager in Washington, DC, wrote in an internal memo in 2005. "It is different from what we told Congress we would do, and it's not restoration."

By 2007 CERP's projected cost had jumped from $7.8 billion to $10 billion, and additional efforts, such as a restoration of the Kissimmee River, were projected to push the cost of restoring the Everglades to as much as $20 billion, according to a report from the Government Accountability Office, a nonpartisan federal agency that investigates how taxpayer dollars are spent. The report also raised fears the work would provide more water for the region's growing population rather than for the ailing river of grass.

One reason CERP progressed slowly was that, although Congress approved the entire plan in 2000, each project within the plan requires additional congressional authorization. More than fifty projects are part of CERP, and because of their size and complexity several include multiple components like reservoirs and stormwater treatment areas. Overall, CERP contains sixty-eight components. Congress approved four projects in 2007, four in 2014, one in 2016, and one in 2018.

A 2012 study commissioned by the Everglades Foundation found Florida's agriculture industry was responsible for three-quarters of the Everglades' phosphorus pollution but had so far paid a quarter of the cleanup costs. The study, produced by RTI International, a North Carolina–based independent nonprofit research institute, found that the public had paid the rest. The "polluter pays" amendment had made polluters "primarily" responsible for cleanup costs. Sugar growers characterized the study as flawed and said they had paid not only to implement best-management practices but also taxes that helped fund the cleanup.

In 2016 new research suggested that historically the Everglades were much wetter than previously thought, prompting the National Academies of Sciences, Engineering, and Medicine to warn that storage capacity might fall short of what was envisioned under CERP by more than 1 million acre-feet. Among the problems was the more than three hundred wells environmentalists had worried about. The amount deemed feasible had been reduced by 60 percent through further study. The scientists also found that the research driving the restoration might be out-of-date, as the CERP evaluations required every five years had not routinely been conducted. The scientists recommended an analysis that would consider climate change and other scientific advances.

Meanwhile less than 20 percent of CERP's total cost was funded, pushing its projected completion beyond the initial target of 2030 to past 2060. This was troubling as ecological problems continued. Toxic algae blooms in 2015, triggered by flows from Lake Okeechobee to the St. Lucie and Caloosahatchee Rivers, prompted emergency declarations in four counties. Seagrass die-offs plagued Florida Bay. High water threatened Lake Okeechobee's aging dike. Everglades marshes drowned under too much water or withered with not enough. By now some project components were completed or nearing completion, but the benefits were measurable only in a small portion of the restoration area—on the periphery of the Everglades, the scientists said.

By 2017 newly restored bends of the Kissimmee River were reviving floodplains and wetlands that would slow the flow and cleanse the water of nutrient pollution from the urban centers and cattle ranches of Central Florida. Construction was under way on reservoirs east and west of Lake Okeechobee, and south of the lake, filter marshes were cleansing the water flowing from the cane fields. Refurbishment of the lake's aging dike was progressing, and new and reconfigured canals were delivering more water to Everglades National Park than ever before. Since 2013 three miles of the Tamiami Trail, which had served essentially as a dike through the heart of the Everglades, had been elevated, enabling the water to flow freely into the Shark River Slough, historically the deepest and wettest part of the Everglades.

By 2019 one CERP project was completed, components of multiple projects were finished, and three projects were in planning. Together the Everglades projects represented one of the world's most substantial efforts at ecological restoration, and with a projected cost of up to $17 billion. The effort, encompassing an 18,000-square-mile region, twice the size of New Jersey, served as a model for other restorations. Work on the Louisiana coastline was modeled so closely after CERP that engineers were sent from Jacksonville to New Orleans to help out. International delegations routinely visit Florida to learn about our water management here. The 50-50 cost share is unique, demonstrating Florida's commitment to the effort. Under most restorations the federal government pays 65 percent, and the state pays 35 percent. More than sixty environmental organizations are involved in Everglades restoration through the Everglades Coalition, and CERP was not even the only restoration being pursued in the region.

Mary Barley remained engaged in the background, relying on the leaders of the Everglades Foundation and Everglades Trust to communicate with the

Mary Barley in Everglades National Park in 2008

Photo by Amy Green

public. She traveled to Washington, DC, two or three times a year to lobby on behalf of the Everglades and Florida Bay.

"You have to have one message and one voice. Otherwise you're message isn't clear," she says. "Every day I'm probably doing something with the Trust."

MANY FLORIDIANS WOULD CONSIDER TOXIC ALGAE to be an obvious ecological indicator of the Everglades—a sure sign something is wrong! The smell alone can be sickening. But for scientists the better indicator of water quality actually are spongy brown mats, each representing little communities of organisms, including fungi, bacteria, animals, and plant detritus, which together form the foundation of the ecosystem. These mats are called periphyton.

Periphyton are little ecosystem powerhouses, transforming the Florida sunshine into energy through photosynthesis. When other animals, such as snails, frogs, aquatic insects, and fish, feed from the mats the energy is passed on, and passed on again as these animals are consumed by larger animals like wading birds. Eventually the energy is spread across the ecosystem.

That is not all. Periphyton produce so much oxygen they are like little floating houseboats for burrowing creatures like worms and insects. As the oxygen dissolves in the water it becomes important for animals with gills, like fish, tadpoles, and aquatic insects, which without the oxygen would suffocate. Perhaps most important to scientists is that periphyton are highly sensitive to phosphorus. Too much phosphorus, and the periphyton will give way to cattails.

Like alligators, periphyton represent another indicator species of the Everglades, and what they, too, are telling us is that thirty years after the federal government sued the state over sugar growers' pollution of the watershed, much work remains to be done. Periphyton nutrient content and biomass show that while phosphorus levels have declined substantially, water quality remains a concern. The measurements show water quality has declined in the Water Conservation Areas since 2014, likely caused by increased water flows and the phosphorus that flows with them. The measurements also show the periphyton have a high sensitivity to phosphorus levels that change with water flow. This explains why wet years show greater impairment in the periphyton than dry years.

IF LAKE OKEECHOBEE IS the Everglades' hydrological heart, the Kissimmee River is perhaps its spine. Historically the river meandered 103 miles to the state's largest lake, with a floodplain spanning up to two miles that supported wetland plants, wading birds, and fish. The Kissimmee was reduced as the Everglades were drained to a 30-foot-deep straightaway called the C-38 canal. In 1992 Congress authorized the river's restoration. Today the effort is scheduled to be complete by 2020 and already is showing early signs of success.

For decades dirt dug for the C-38 remained heaped on its banks. Bulldozers pushed the dirt back into the waterway, filling it and making way for the river's old meanders to recarve their historic path. Five dams controlling the flow were blown up, enabling the river's natural flow.

When the $75 million restoration is completed more than 40 square miles will be restored, including 12,398 acres of wetlands and 40 miles of historic river channel. Already wetland plants are thriving, including pickerelweed, arrowhead, Carolina willow, and buttonbush. Bird populations are rebounding, including those of wading birds like white ibis, great egret, snowy egret, and little blue heron. In some years the increase has been double the expectations. Ducks have returned, such as the American widgeon, northern pintail, northern shoveler, ring-necked duck, and black-bellied whistling duck. Eight shorebird species that were absent before the restoration are back, including breeding black-necked stilts. Dissolved oxygen has increased sixfold, and largemouth bass and sunfishes now comprise 63 percent of the fish, up from 38 percent.

The US Army Corps of Engineers touts the Kissimmee's rebirth as one of its most successful restoration efforts, showing that Mother Nature's response time to a reclaimed habitat is rapid. The success perhaps serves as a ray of hope for the Comprehensive Everglades Restoration Plan, even as the effort is a long way from completion.

18

THE RESERVOIR

U.S. Sugar Corp. executives stunned everyone back in 2008 by agreeing to sell all of the company's land to the state for Everglades restoration. Eight years later they stunned everyone again by changing their minds, basically, and taking up another bitter battle *against* a proposal to put some of the land toward a reservoir south of Lake Okeechobee.

In 2016 a proposal emerged for the reservoir that everyone had already agreed to as part of the Comprehensive Everglades Restoration Plan. In fact, the reservoir would be the same project under CERP that workers had stopped construction on in 2008 when Governor Charlie Crist announced the state would buy out U.S. Sugar. Since then, the unfinished reservoir had been repurposed into a shallower reservoir or basin, used during times of high water to treat the water for phosphorus. (In 2017 the basin opened to the public for hiking, biking, and seasonal alligator hunting.) Now environmental groups wanted to get construction going again on the CERP reservoir project that had been abandoned in 2008—a large reservoir in the Everglades Agricultural Area.

State senate president Joe Negron introduced the proposal, in which the state would buy 60,000 acres south of the lake for the reservoir. Negron, a Republican, represented Stuart, a coastal community east of Lake Okeechobee near where the St. Lucie River, Indian River Lagoon, and Atlantic Ocean converge. Toxic algae infested the waters here in 2016, and people had had enough. Crist's deal with U.S. Sugar had been downsized during the recession, but the state still retained an option to acquire the rest of the land by 2020. The senate president said he would make the reservoir, aimed at stemming the toxic flow, his "No. 1 personal priority."

THE RAINS CAME EARLY IN 2016. Normally the wet season does not arrive in South Florida until May, but that year the rains began in January, and Lake Okeechobee rose rapidly. When flows of water west to the Caloosahatchee River failed to bring down the lake fast enough, the US Army Corps of Engineers discharged water east to the St. Lucie River.

By February the Big O continued to rise, and the engineers rushed to dump as much water as possible in all directions. The water conservation areas topped out, and more than 5.6 billion gallons were moved into Everglades National Park after Governor Rick Scott asked the corps of engineers to speed up a restoration project. The South Florida Water Management District was pumping up to 96 billion gallons daily into an emergency detention basin near Miami International Airport constructed after Hurricane Irene in 2000 and was dumping more water into the Miami Canal. The water managers also were pumping another 5 million gallons daily into an aquifer storage and recovery pilot well in Palm Beach County.

Many hoped some of the freshwater might find its way through South Florida's maze of water management infrastructure to Florida Bay, which was reeling after a severe summer drought the year before had caused its salinity to soar. During the following months a seagrass die-off had spread to more than 62 square miles, encompassing some of the same areas affected in 1987 by the problems that had incited George Barley into action. Now there were fears a similar sequence of events was unfolding.

"Because we had so much rain it should have been plenty of water to provide to the bay," one biologist said at the time. "It points to the inability of the water management district infrastructure to supply Florida Bay with its needed water."

Mary Barley's Everglades Trust took out full-page ads in March 2016 in the *New York Times* and *Miami Herald*. The ads featured a letter written by Mary to William White, chair and chief executive officer of the Charles Stuart Mott Foundation and board member of U.S. Sugar. The foundation held a board seat at the Environmental Grantmakers Association, meeting that week in Miami. The letter read:

> Dear Mr. White:
>
> Try as we have over the years, Everglades advocates have been unable to secure a meeting with you to discuss the Mott Foundation's unique opportunity to assist in restoring and preserving the Everglades—an inestimable national resource that is essential to Florida's economy, health, and recreational vitality.
>
> The Mott Foundation claims it is committed to preserving freshwater ecosystems as part of its espoused mission "to maintain an ethic of respect, integrity and responsibility," while recognizing that "each individual's quality of life is connected to the well-being of the community, both locally and globally." I couldn't have said it better myself, especially when considering that the Everglades is one of the world's largest, most complex freshwater ecosystems—the primary source of drinking water for the nation's third most populous state and home to more than 70 threatened and endangered species, including the Florida panther. Given the recent crisis with contaminated drinking water in your hometown of Flint, Michigan, surely you understand the importance of protecting water supplies.
>
> Despite the Mott Foundation's publicly avowed commitment to preserving and protecting freshwater ecosystems, we remain deeply disappointed that U.S. Sugar Corp.—a company that is a major asset and source of income of the Mott Foundation—is a steadfast opponent of science-based public policies to restore the Everglades and is one of the largest sources of pollution to

the Everglades. In fact, the company was convicted of knowingly dumping hazardous waste that went into the Everglades.*

It is baffling and ironic that the Mott Foundation has demonstrated such a deep commitment to the environment and preserving freshwater ecosystems outside of Florida while controlling a company that is integral to the destruction of one of the most important freshwater ecosystems in the nation.

It is hardly an overstatement, Mr. White, to say that you have an extraordinary opportunity to significantly advance the restoration of the Everglades, given your leadership of both the Mott Foundation and U.S. Sugar Corp.

Restoring America's Everglades, which means so much to so many, depends upon U.S. Sugar Corp.'s participation as a full and committed partner, as opposed to it remaining a steadfastly uncommunicative, unwavering obstructionist.

We hope you will change your position and meet with me soon. I look forward to hearing from you.

Sincerely,
Mary Barley
President, The Everglades Trust

A U.S. Sugar spokeswoman responded at the time that growers had cut pollution in the Everglades. She described the Everglades Trust as Barley's "radical friends" and said the company's leadership had met with the trust "numerous times over the years and that farmers in the Everglades Agricultural Area had given up nearly 120,000 acres to assist with restoration."

THE TOXIC ALGAE BEGAN TO BLOOM in May 2016, first in Lake Okeechobee and then in the estuaries as hundreds of billions of gallons flowed from the lake. Normally the blue-green algae, or cyanobacteria, cannot survive

---

* *U.S.A. v. United States Sugar Corporation*, Case No. 91-8125-CR-PAINE.

in saltwater, but water managers were dumping so much freshwater that the algae forced Atlantic beaches to close over July Fourth weekend. Environmental advocates described the blooms as the worst they had ever seen. Fishing guides reported business was off by as much as 50 percent. In Lake Okeechobee the toxicity was 200 times what the World Health Organization considered safe. Governor Rick Scott declared a state of emergency.

Among the hardest-hit businesses was Central Marine in Stuart. The marina is located in an alcove where the water experiences little circulation, enabling vast clumps of the foul-smelling guacamole-looking algae to bloom, then die and rot all in one place, layer upon terrible layer. Employees coped with face masks, air purifiers, and air fresheners.

"You can see the flies that are on the top of it. They're eating the rot, so that's like the sewage that is out there. You can see the big brown spots that look like sewage," the marina's manager told me at the time.

Another employee said, "If you took a Porta Potty that's been out at a music festival all weekend and filled it with dead animals and then shook it up in a blender and then left it in the sun, that's probably what it smells like."

"Good days," another remarked, "it smells kind of like, it smells like rotten meat and cat litter. The bad days it smells like a septic tank backed up into a pig farm."

In November 2016 Mary published the following opinion piece in the *Miami Herald*:

> We heard a lot this political season about government corruption. While not all corruption is illegal, every instance of it is certainly unethical. Americans just showed they are willing to take extreme measures to combat corruption, and our local officials would do well to take heed, for a dangerous corruption is on open display here in Florida, and by any means legal and possible we will eradicate it. Indeed, we must.
>
> In Tallahassee, Gov. Rick Scott subverts the public interest and does the bidding of Big Sugar in exchange for campaign cash. Big Sugar, in this instance, is U.S. Sugar Corp., whose president and CEO is Robert Buker, Jr., and Florida Crystals,

owned and operated by the Fanjul family. The result is that the Everglades are in peril and therefore so, too, is the future of South Florida and its residents.

In order to survive, the Everglades needs a steady supply of fresh water from Lake Okeechobee. But even if that water could reach the Everglades and ultimately the Florida Keys, it is heavily polluted with nitrogen and phosphorus from agricultural runoff. That's the reason past summers have seen such toxic waters and algae blooms along the coasts, decimating the fishing, small businesses and tourism.

Incidentally, much of the pollution in Lake Okeechobee is from previous decades of Big Sugar backpumping toxic water from its land to the south directly into the lake. Such polluting is routine; U.S. Sugar Corp. has been convicted of knowingly dumping hazardous wastes into the Everglades.

There is a solution that is incredibly simple and scientifically sound. Water from Lake Okeechobee can be stored in land just south of the lake, where it can be cleaned, keeping toxic water out of the St. Lucie, Caloosahatchee, Atlantic and Gulf and providing water for the Everglades and the Florida Keys in the dry season and staving off saltwater intrusion. It could also provide South Florida all the drinking water it needs. Again, very simple and backed by hard science.

Enter politics.

Big Sugar owns that land just south of the lake. In 2010, U.S. Sugar signed a contract finalizing the sale of up to 153,000 acres by October 2020. As for the money to buy it, the voters of Florida indicated where that should come from—themselves. In the election of 2014, 75 percent of them, an astonishing majority in politics, approved the Florida Water and Land Conservation Amendment, which provided the state the money it needed to complete this purchase, among other projects.

But U.S. Sugar decided, in the end, it didn't really want to sell. How was it so easily able to thwart the overwhelming will of the people? Money. And lots of it. In the past 10 years, Big Sugar has given upwards of $60 million and more just for statewide elections. That doesn't count federal elections, and even that number is likely to be too low since the Citizens United ruling has made it easy for corporations to hide just how much money they're donating to politics.

It is clear why Big Sugar does not want to sell this land: greed. It really does come down to more big houses and more shiny cars for a handful of them at the expense of drinking water, a healthy Everglades, and safe oceans for millions of us. That greed is so short-sighted, for it's their drinking water, their Everglades and their oceans, too.

Less clear is Gov. Scott's lack of courage to uphold his duty to defend Florida's Constitution and enact the will of its people, unless he's planning a further career in politics, for which there is that unending need of campaign money. But if he continues to bend to the will of Big Sugar, it is not hyperbole to say that the fate of South Florida and, therefore, Florida, hangs in the balance.

But again, there's another simple solution. In 2000, Congress passed the Comprehensive Everglades Restoration Plan (CERP), part of which called for Big Sugar to sell the acreage below Lake Okeechobee needed for restoration. Big Sugar is on record as supporting CERP. Some of its top officials were active participants in the meetings that created the legislation. All it would take is for Big Sugar to identify 60,000 acres it would sell to the state as a reservoir below Lake Okeechobee. Failing that, the governor could put public interest ahead of private gain and use eminent domain to acquire those acres.

The solution is simple; the way forward clear. As citizens we are stewards, and if we shirk that responsibility, this land of South Florida that we call home will someday be unable to sustain us, all of us, including Governor Scott and those rare few atop Big Sugar. Their deed can become our salvation as we work truly together to help our neighbors and preserve our unique natural heritage.

WHEN THE STATE LEGISLATIVE SESSION began in January 2017 lawmakers took up senate president Joe Negron's proposal for a reservoir south of Lake Okeechobee. The reservoir was aimed at stemming the toxic algae by moving more water south rather than east and west. The measure before lawmakers would speed up the reservoir's construction by directing the South Florida Water Management District to start the planning right away. The water management district was to acquire the 60,000 acres from willing sellers, but if there were none, the district was to exercise the last remaining option under the state's land-buy deal with U.S. Sugar.

Sugar growers lined up against the proposal. They characterized the reservoir as scientifically unsound and unnecessary, even though everyone had already agreed to the reservoir under CERP. The growers favored a reservoir north of Lake Okeechobee, also part of CERP, and argued that the acquisition of more farmland represented an economic threat. One U.S. Sugar spokesperson went so far as to say a thousand jobs were in jeopardy, although company executives hadn't seemed too worried about that back in 2008 when they agreed to sell all of the company's land to the state for Everglades restoration. The South Florida Water Management District wanted to stick to a new schedule, one that had construction starting in 2021, even though the US Army Corps of Engineers said the schedule could be adjusted and construction had already been under way back in 2008 when Crist announced the state would buy out U.S. Sugar. For Mary Barley it was exasperating.

"Their whole basis is a lie, and so they throw something out there like that. It's not based on science, and they know it," she says. "You just have to stick with the facts. The foundation is science-based.... You stick with the message, and the message is about clean water, which means healthy citizens and a healthy environment."

When I traveled to the Everglades Agricultural Area in early 2017 to report on the issue, at the heart of much of the debate was a 2015 University of Florida (UF) study examining how to spare the estuaries by moving more water south rather than east and west. I encounter this all the time as an environmental journalist: everyone agrees we should do what the science says, but no one agrees on what the science says. The UF water study, as everyone called it, was clear. It said more storage was needed north and south of the lake and that existing and currently authorized projects were not enough. The study said significant long-term investment would be needed to improve the region's water management infrastructure. It recommended the following storage volumes:

- 400,000 acre-feet within the Caloosahatchee River watershed
- 200,000 acre-feet within the St. Lucie River watershed
- 1 million acre-feet north and south of Lake Okeechobee

Farmer Rick Roth says the reason sugar growers opposed the reservoir was that the proposal called for it to be built on private land. The earlier reservoir had been under construction on land the state already had acquired from the Talisman Sugar Corp.

"I hate to be controversial, but part of the thought process of building the EAA [Everglades Agricultural Area] reservoir was to intentionally take land out of production, and that should not be the goal," he says. "The goal should be to clean up the water, not to take land out of production."

For Mary the debate grew personal as two of George Barley's daughters had opposing op-eds published on the issue. In March 2017 Catherine Barley-Albertini, now living in Cardiff-by-the-Sea, California, had the following op-ed published in the *Orlando Sentinel*:

> My father, George Barley, never ceased to amaze me with his passion for nature. He took my mother, Shirley, and sisters Lauren and Mary and me out in it often, and our family vacations centered around it. He taught us to love and protect nature and wildlife.
>
> Although I live in California, I visit Florida frequently and appreciate its delicate ecosystem. If my father were alive, he would be appalled at today's political, economic and scientific shenanigans. Almost 22 years after his death, nothing has really changed, despite people at his funeral promising to fight for his cause. He was wealthy, but he was an extremely conservative, low-key person who lived modestly. More than anything else, he demanded honesty and stopped at nothing to find the truth. He instilled that precious trait in me. While he wasn't able to realize his dream, I feel an obligation to help fulfill it.

He never wavered in his pursuit of mitigating environmental damage to the Everglades. Protecting and restoring the "River of Grass" and his beloved Florida Bay were goals he relentlessly pursued, even as he died while traveling to meet the Corps of Engineers that fateful day. Still greatly respected today, his name is nearly synonymous with the historic Everglades restoration that has been underway for nearly three decades.

Sadly, his dream of saving the Everglades is slipping away, as that focus has been replaced by the battle pitting coastal environmental groups against agriculture over damaging Lake Okeechobee discharges.

Much to our dismay, current environmentalism is just another special interest relying on glitzy galas, well-heeled lobbyists and paid staffers to spread its messages. Where my father used his passion to urge the public and private sectors and our political leaders to come together to take action, today's activists are spreading a message of hate and division.

My father's battle with farmers was to ensure that those responsible for pollution paid their fair share. He fought for stricter water-quality requirements and for farmers helping to restore the Everglades. He'd be happy that sugar-cane farmers are cleaning their water and paying their fair share of restoration. Mostly, he'd have been thrilled with the incredible progress today with more natural water flow and more than 90 percent of the Everglades meeting strict clean-water standards.

Despite this progress, today's Everglades Foundation, the organization that he founded, has strayed far from my father's mission. False science is again pushing false solutions. I've conducted extensive research and talked to many scientists regarding the extensive flooding of Florida Bay with excess nitrogen in the 1990s—creating massive algae blooms and wildlife mortality. In this my father was misled by scientists Jay Zieman and Ron Jones. Now folks are callously dismissing the generational family farming communities south of Lake Okeechobee, calling to flood their land and the Everglades.

I don't know where my father would stand on Lake Okeechobee discharges, but I know he was focused on keeping the Everglades from further harm. My father would never support a plan to send massive amounts of polluted lake water south to the Everglades when it was already too full. He would consider the issue more comprehensively, balancing the entire ecosystem, north, south and central, while considering the complex and comprehensive effects of the many septic systems as well as the effects of nitrogen, fertilizers, pollution and pesticides from our air and soil.

He would have hired experienced and accurate scientists and weighed their opinions in making a decision. When it comes to humans, flora and fauna, there is no room for error.

Getting the proper timing, quantity and quality and more natural conveyance of water south has been the focus of literally billions of dollars of state and federal monies. With development, the remaining Everglades is half its original size and rerouting hundreds of billions of gallons of nutrient-rich Lake Okeechobee water south has never been a part of Everglades restoration. It sounds simple, but sending that much additional lake water south would destroy what's left of the Everglades.

The Everglades Foundation that my father founded has badly lost its way under the current leadership, including my stepmother and board member Mary Barley and co-founder and board member Paul Tudor Jones. Rather than keeping momentum for Everglades restoration moving forward and ensuring funding is not lost to other priorities, they have abandoned my father's dreams for their agenda.

Damaging discharges to the estuaries need to be addressed, and we need to keep that water out of the lake and out of the estuaries. However, the solution to those problems cannot come at the expense of the recovering Everglades.

The current debate playing out in the Florida Legislature must stay focused on solutions that continue real restoration. We have only one Everglades.

A few days later Mary Barley Hurley, another of George's daughters, now living in New Smyrna Beach, Florida, had the following op-ed published in the same newspaper, the *Orlando Sentinel*:

> The Everglades continues to need champions right now, not bad politics as usual or irrelevant distractions.
>
> It appears that Big Sugar and its highly paid public-relations firm are conducting business as usual by trying to distract and capitalize on family dynamics rather than address the issue of greater importance, which is, and always has been, the restoration of the Everglades.
>
> I am the youngest daughter of George McKim Barley, and I am writing this in response to a guest column written by my sister Catherine Barley Albertini, in which she states that our father would be opposed to the restoration plan that has been worked on diligently by people who were chosen because they were like-minded to my fathers' vision.
>
> My stepmother, Mary Barley, has devoted her life to this cause after my father's passing and regardless of anything that goes on in families, I am confident that she, along with Paul Tudor Jones and many others, have maintained his original vision and have even made it bigger and stronger, never wavering.
>
> The column suggested my father "would never support" the reservoir south of Lake Okeechobee that has been proposed by Senate President Joe Negron. Nothing could be further from the truth. In fact, this reservoir is exactly what my father worked for, and it's one that I fully support.
>
> The column published in the Orlando Sentinel, which has since been circulated by the sugar industry's British-owned public-relations firm, parroted the talking points of the very industry that polluted the Everglades and is now opposing the reservoir—just as it has fought virtually every effort to clean up its own mess. This latest stunt is just more smoke and mirrors and more stalling.
>
> The fact that two-thirds of the fresh water that once flowed into Florida Bay had been siphoned off, killing seagrass and

the wonderful species of sport fish that spawned and fed on it, was heartbreaking to my father. He believed the discharges of excess, polluted storm water into our coastal waterways was a crime against nature.

Those new to Florida might be tempted to think these discharges and the algae outbreaks they cause along both of Florida's central coasts are a new thing, but sadly, they are not.

For decades, the Army Corps of Engineers has been forced to discharge billions of gallons of polluted Lake Okeechobee water into the Caloosahatchee and St. Lucie rivers and estuaries. This has tremendously upset the delicate balance of saltwater and fresh water, causing the recent and tragic outbreaks of algae that are toxic to wildlife and humans alike.

My father fought with every ounce of his formidable energy to end this wasteful and destructive practice. He knew the health of the Everglades and his beloved Florida Bay depended on restoring the natural southerly flow of clean fresh water. The practice of channeling water and flushing the excess out to sea to accommodate Big Sugar has to stop and the Everglades Agricultural Area reservoir is critical to replace this reckless and damaging practice.

At the time of my father's premature and untimely death, he was working both in Washington and Tallahassee to ensure passage of the legislation that ultimately became known as the Comprehensive Everglades Restoration Plan. Due in large part to my father's own efforts, the second project identified in CERP legislation—out of 68 projects—was the very southern reservoir that the Florida Legislature is now about to approve.

The legislation my dad worked so hard to pass envisioned massive increases in water storage on all sides of Lake Okeechobee, primarily in the South. The sustainability and preservation of the water supply for 8 million people in South Florida can be accomplished through this, as well as the restoration of the Everglades and Florida Bay.

> To the south of the reservoir, the plan projected the construction of man-made wetlands that would clean the water before being released into the Everglades and Florida Bay.
>
> Today, those wetlands are nearing completion, and I know my father would be so happy to see that this is being accomplished.
>
> This is only phase one, though. It is now imperative that we proceed with the reservoir that will make this a reality. The southern reservoir in the EAA will be the completion of the journey that my father so voraciously and courageously fought for. And he fought so hard for it because he knew it was the right plan.
>
> I did not pen this opinion for my stepmother and the Everglades Foundation or against my sister, but because I know it's what my dad would have wanted.

Later that spring Negron revised his proposal for the reservoir south of Lake Okeechobee. The new plan relied on state land for at least the first phase of construction but left open the possibility that land could be acquired from willing sellers in the future. The plan prohibited eminent domain. In May 2017 lawmakers approved the plan.

Kimberly Mitchell, executive director of the Everglades Trust, had the following op-ed published that month in the *Miami Herald*:

> When George Barley began the effort to restore the flow of clean freshwater to the Everglades and Florida Bay in the 1990s, he was told that the best government could (or would) do was clean out the culverts under Tamiami Trail.
>
> For the next two decades, the battle to save the Everglades has been mostly a David vs. Goliath effort.
>
> Goliath was Big Sugar: Florida Crystals, U.S. Sugar and their armies of lawyers, lobbyists and politicians.
>
> David was the group of scientists, hydrologists, wetland experts, engineers and citizens who knew that Big Sugar was polluting the Everglades and blocking the natural southerly flow of water from Lake Okeechobee.

When the history of Everglades restoration is written, 2016 will be remembered as a pivotal point. As record rains during 2015–2016 tested the capacity of Lake Okeechobee, the Army Corps had no choice but to send the excess into the Caloosahatchee and St. Lucie rivers, setting off an environmental catastrophe: historic algae blooms that forced the affected areas to endure a "State of Emergency" that lasted 242 days.

As TV beamed images of slimy blue-green algae and residents complained that it smelled like "death on a cracker," Big Sugar stood like a Colossus to block a solution. One argument after another was tossed into the public dialogue. As one legislator observed, their arguments were like a game of "Whack-a-Mole": as soon as one spurious argument was refuted, another was advanced.

For years, that strategy had worked, but this year, as NASA's satellites showed the 239-mile algae bloom was clearly visible from space, the destruction became obvious even within the Tallahassee bubble.

The lost summer of 2016 led citizens to organize. New groups named Bullsugar and Captains for Clean Water stood shoulder-to-shoulder with older organizations like the Everglades Trust, Bonefish Tarpon & Trust, Florida Sportsman Magazine and the Everglades Foundation. Soon, corporate giants like Orvis, Patagonia, Costa Sunglasses, Guy Harvey, and others joined the effort.

Soon, Florida's political leaders got on board, signing the "NowOrNeverglades Declaration" and earning the endorsements of the grass roots. Nearly 100,000 Floridians signed on: they called, they emailed and they met in person with their legislators. They spoke as one to demand action and used social media to engage with like-minded people across the state.

Armed with overwhelming evidence and science and backed by tens of thousands of supporters, Senate President Joe Negron made the EAA Reservoir his top legislative priority. Even at great personal sacrifice—he resigned his lucrative law partnership to wage this fight—Negron showed extraordinary integrity in leadership.

Meanwhile, the "Now Or Neverglades" movement grew louder and stronger. Boat captains, fishing guides, realtors, rowing instructors, college students from all parts of Florida, hoteliers, Chambers of Commerce, moms and dads and their children, small business owners and health care professionals made repeated trips to Tallahassee testifying to the legislature in gut-wrenching terms about the nightmare they were being forced to endure. They begged the lawmakers to adopt the Negron plan.

In response, Big Sugar cranked up its PR machine, blasting out newspaper columns that they penned for their sympathizers. They spent thousands on daily TV commercials proclaiming their corporate citizenship and the "food" they produce.

They advanced a ludicrous, desperate argument that their fellow citizens were "anti-farmer" and hired 100 lobbyists and lawyers to quash the southern reservoir. They launched and underwrote faux "citizens" groups, like "Floridians for Clean Water" and "Stand Up North Florida" in a losing effort to confuse the public.

Bold and steadfast leadership in the Florida Senate from Joe Negron, Jack Latvala and Rob Bradley proved to be too much for the once all-powerful sugar lobby. The Florida House, led by Richard Corcoran, Thad Altman and Heather Fitzenhagen, agreed to the plan.

In the end, the good guys finally won. "Now Or Neverglades" became the unstoppable force that overcame the immovable object, and we are finally on our way to sending clean water south.

Somewhere, George Barley must be smiling.

THE MEASURE LAWMAKERS APPROVED also called for the South Florida Water Management District to terminate the state's last remaining option to U.S. Sugar land. In December 2018 the district's board members quietly voted to terminate the option. With that, the deal that had been touted as a historic breakthrough in Everglades restoration was dead.

The Everglades Foundation called on the board members to hold off on their vote until newly elected governor Ron DeSantis could take office. US representative Brian Mast, a Republican representing the same algae-fatigued area as Negron, demanded that the entire board resign.

It would be among the first skirmishes between the South Florida Water Management District, the state agency overseeing Everglades restoration, and DeSantis, who had campaigned on a platform focused in part on toxic algae and the Everglades.

19

WHAT IS RESTORATION?

Hold in your head an image of a fully restored Everglades. Look around. Examine the water, landscape, wildlife. What do you see? For Mary Barley, the image is one of Florida Bay.

"If Florida Bay is healthy, we are like the canary in the coal mine. To me if it's getting enough water and it's clean, then the rest of the system should be functioning as it's supposed to," she says. "It has clean water. The birds are back. The fish species have come back, and there is a healthy amount of recreational fishing. Florida Bay is mainly for the critters, but if Florida Bay is healthy, then that means there is freshwater flowing from Miami, from Fort Lauderdale, from West Palm Beach, and the Biscayne aquifer is being recharged."

Adds farmer Rick Roth: "You have more land for stormwater treatment areas, and you have agriculture continuing to thrive. And so you have all the components that you want to have. You want to have a healthy ecosystem with more land being used for improving the water quality. You have a healthy urban environment, and you have agriculture. It's a very balanced approach."

By now we have discussed why Everglades restoration is needed, traced the effort's path from concept to construction and considered how much it all might cost. Now let's examine the word itself, *restoration*. Often we use the word to describe the process of rescuing something from a dilapidated present and returning it to a gleaming past. We might *restore* an old home, for instance, although in this case we do not use the word in a literal sense, do we? It would be more appropriate to say that we *adapt* the home to fit our present needs by, say, modernizing its electrical work and plumbing.

Similarly, when we talk of Everglades restoration we have to acknowledge it is not possible to return the watershed to its predrainage state a century ago, and most of us would not want to do that anyway. There is too much development here today and millions of people living here, and most of us are happy with modern conveniences like water supply and flood control.

This vision of an untouched Everglades is not the mission of the Comprehensive Everglades Restoration Plan. Instead the effort aims to bring back key attributes and functions like a more natural flow, mosaic landscape, robust wildlife, and secure water supply, all while adapting the watershed to meet the present needs of the region's booming population.

By the time all of the work is completed, beyond the lifetimes of many readers here, taxpayers likely will have paid tens of billions of dollars. What will we be getting for our money?

IN 2010 THE EVERGLADES FOUNDATION began a process of consolidating an ever-expanding body of very technical research into a single document that would be readable for decision makers and the public. A team of scientists, including ones representing government agencies, academic institutions, and the private sector, asked decision makers and resource managers what were their biggest questions about the restoration. The US Department of the Interior funded the study.

Then to answer the questions the scientists modeled five different restoration scenarios that varied in how much and what kind of water storage they offered. One scenario considered a future with no restoration, another with full CERP implementation, and three others with varying degrees of storage. The scientists focused on storage because that is among the hardest problems of restoration: How do we give the river of grass the water it needs while also sparing estuaries to the east and west of Lake Okeechobee of harmful influxes? Complicating the situation is the centerpiece of CERP's storage strategy: 333 aquifer storage and recovery (ASR) wells where the water would be pumped underground. The strategy had always been controversial, and by 2015 a report from the US Army Corps of Engineers and South Florida Water Management District recommended scaling back the number of wells to as few as 131.

After completing their modeling the scientists concluded that some 2.2 million acre-feet of new storage is needed to restore the Everglades. That would represent enough water to flood the entire state of Florida a half-inch deep. To meet the state phosphorus standard of 10 parts per billion the scientists recommended expanding the stormwater treatment areas or filter marshes south of the Everglades Agricultural Area, as they said the current design was for the existing flow from the cane fields, and the hope with restoration was that the flow would increase.

With no restoration, the scientists described a bleak future for the Everglades, with 90 percent of wading birds lost and the Everglades snail kite nearly extinct. The Everglades snail kite is a bird with such a particular taste for the apple snail, its beak is actually curved to help the bird extract the snail from its spiraled shell. Elsewhere, harmful flows to the estuaries would continue while the flow south of Lake Okeechobee would be too wet or too dry. Fish densities would decline, 20 percent of the marshes would be polluted with phosphorus, and Florida Bay would be too salty.

With full CERP implementation more than 2.4 million acre-feet of water would be stored underground, and more than half of the barriers impeding the river of grass's historic flow would be removed. The flow would improve, and phosphorus pollution would decline by as much as 65 percent in the hardest-hit areas. Wading bird and fish populations would increase, and Florida Bay would experience a 50 percent reduction in salinity.

Considering the concerns with ASR wells, the scientists modeled scenarios with no wells and found that expanding surface storage beyond what is in CERP could be as beneficial for the Everglades or better, although they

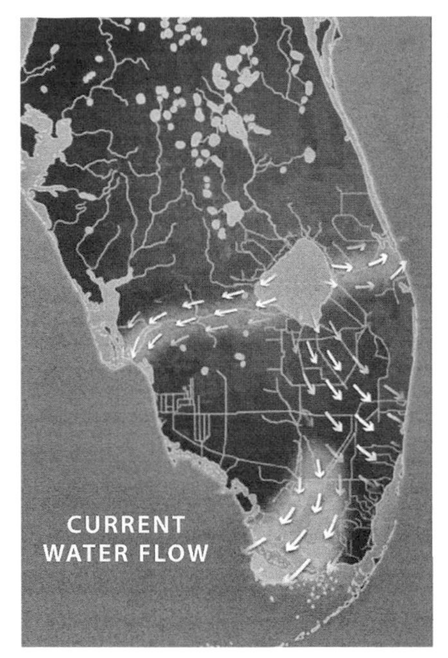

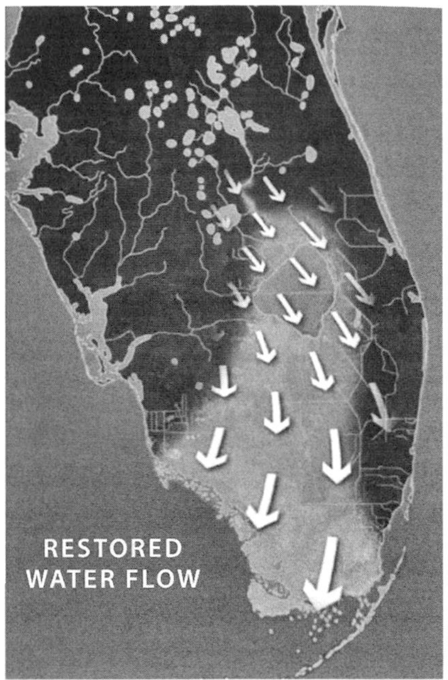

Comparison of the current water flow and the planned restored water flow
Courtesy of the Everglades Foundation

acknowledged this option is limited by geography. The scientists said this would also require removing more barriers impeding flow, but that measure would benefit habitat. When 75 percent of barriers were removed the flow improved, and wading bird and fish densities increased. Phosphorus pollution was reduced by as much as 61 percent in the hardest-hit areas, and Florida Bay experienced a 50 percent reduction in salinity. These options also cost about the same as CERP, although one option cost $4 billion more.

The scientists also concluded there was enough land in Everglades National Park, Big Cypress National Preserve, and the water conservation areas to bring back key aspects and functions of the historic river of grass. They described an Everglades that would be wetter than before, with more alligators and wading birds. They said it was possible the higher water would displace some species accustomed to drier conditions like white-tailed deer, which might find a new home among artificial or restored tree islands. In

George Barley's beloved Florida Bay all of the restoration models showed better salinity, which would help the tarpon he loved to fish for, as well as imperiled species such as the Roseate spoonbill and American crocodile.

Meanwhile, unfortunately a new problem was developing with CERP: the plan was based on a predrainage vision of the Everglades and the assumption that past rainfall and temperatures would continue. But the watershed was changing, especially its climate.

BENEATH SOME OF THE WATER AND SAWGRASS of the Everglades is peat soil. Here, with little oxygen, microorganisms cannot decompose plant remains quickly, and slowly over hundreds of thousands of years the remains accumulate, forming a soil that when fossilized actually turns to coal and burns. Historically, some of the peat piled up to relatively higher elevations, supporting tree islands and sawgrass ridges. Elsewhere the sheet flow maintained the peat at lower levels, resulting in longer flood periods and a deeper slough habitat. Together these topographic highs and lows created a unique ridge-and-slough landscape pattern in the Everglades.

The accumulated peat represents a rich reservoir of carbon and nutrients such as phosphorus and nitrogen and also small amounts of other elements like mercury. Draining the Everglades exposed the soil to conditions that were too dry and also to intense fire, resulting in the loss of vast areas of peat. This flattened the Everglades and caused the release of carbon into the atmosphere, and phosphorus and nitrogen into the watershed. Some marshes of Everglades National Park have lost as much as 3 feet of elevation since the watershed was drained. In places where farming-related phosphorus has replaced the sawgrass with cattails, the soil is different. It is not peat and not a desirable restoration condition. The phosphorus in this soil is virtually impossible to remove and stands to harm the Everglades far into the future.

Peat collapse in the Everglades was first documented in Cape Sable in 1920 when newly dug canals brought saltwater into freshwater marshes. This caused a dramatic transformation of the habitat in only ten or twenty years. When saltwater is exposed to freshwater marshes it causes a cascade of events that lead to, basically, a hole in the landscape. Freshwater plants become stressed or die back, and this slows the production of roots that contribute to peat formation. Bacterial activity increases, and the peat breaks down. Eventually the marsh collapses.

Open water replaces these collapsed marshes, leading to a net loss of land similar to what has been documented in Louisiana, although the rate of elevation loss here (as much as a half foot or more a decade) appears to be more dramatic. The carbon released into the atmosphere can exacerbate climate change, and the nutrients can flow into Florida Bay and vulnerable coastal estuaries, where they can trigger harmful algae blooms and other problems.

By 2018 it was clear that the Everglades were facing a new threat. Sea level rise of some eight inches during the past century, combined with more frequent and intense hurricanes and less freshwater flow from the north, were contributing to an almost literal collapse of the watershed in some places. Elsewhere mangroves, small trees that thrive in coastal brackish waters, were moving inland by the hundreds of thousands, replacing the sawgrass. The saltwater was creeping into the underground aquifer, threatening the region's drinking water supply.

Rainfall and temperatures were also deviating from the historical patterns on which the Comprehensive Everglades Restoration Plan was based, prompting the National Academies of Sciences, Engineering, and Medicine to call in their latest report for a "mid-course assessment" of the work to account for climate change, an issue the scientists said had been neglected in the planning up to that point. The scientists pointed out the effort's slow pace, calculating that if recent federal funding levels continued and were matched by the state, construction of the rest of the congressionally authorized projects could take as long as sixty-five years, putting the end date well beyond 2080. The scientists emphasized that restoration leaders needed to plan for the Everglades of the future. The report said that while recent planning had brought the vision of water storage under CERP into sharper focus, a more holistic understanding was needed of how the CERP projects would function together and also how they would stand up in a warming world.

Across the globe climate change represents a challenge for restoration efforts, from wetland preservation in Louisiana to river management and flood control in the Netherlands. The problem is acute in the Everglades because the restoration work is especially complex and costly. One solution is an approach called adaptive management, which allows for adjustments to the work as environmental conditions evolve. In the Netherlands, for instance, plans might call for temporary storage areas, rather than dikes, that can handle only a certain amount of water, potentially creating more capacity for the future. The problem in the Everglades is that the water storage does not even meet immediate needs, much less allow for change in the future.

The consensus among the scientific community is that Everglades restoration will make the watershed more resilient, as more freshwater beats back the encroaching saltwater. This would reduce peat oxidation and marsh collapse and allow for more water management flexibility.

WHEN I TRY TO IMAGINE how George Barley might have envisioned a restored Everglades, my mind can't help but settle on a 4-inch white binder labeled on the front cover with the single word CONFIDENTIAL. The enormous binder is dirty, its pages weathered with age and mildewy from years spent on a shelf of Mary Barley's garage in beachy Islamorada, with other personal records of the Everglades power couple's tumultuous years in politics.

The binder contains hundreds, perhaps thousands, of pages documenting George's work on behalf of Florida Bay and the Everglades. The six-page "Florida Bay Chronology" begins with the Central and Southern Florida Project for Flood Control and Other Purposes and draining of the Everglades. The chronology concludes with the 1994 Everglades Forever Act. The binder also contains the letters scientists sent about their concerns for Florida Bay to George as chairman of the Florida Keys National Marine Sanctuary Advisory Council. Also there are government documents and correspondence George amassed that he felt proved sugar growers were not paying their fair share of Everglades cleanup, as well as the existence of a government "cover-up" in the early 1990s aimed at concealing from the public the true cost of the restoration. George used the binder to help prepare for his *60 Minutes* interview.

I thought *I* was obsessed with the Everglades. But George spent an untold number of hours during the last years of his life going to government meetings and exchanging faxes and letters with the bureaucrats and politicians who were shaping the historic restoration in the early years—all part of a mission to make sure Florida Bay and the Everglades got saved and the appropriate people paid the cost. To say he dedicated this phase of his life to the cause would be more than appropriate.

"SFWMD [South Florida Water Management District] said, in July 1993, that this project would cost $465 million," he wrote at one point. "I suspected they were covering up the true cost of the project, and started making inquiries. I felt it would be hard to dig these costs out if the inquiry was too obvious, so I decided to look into the costs of the Everglades Nutrient Removal (ENR) project, a completed 3600 acre SFWMD test treatment pond designed to research treatment of sugar's pollution. This project would be incorporated into the announced overall project, so the twofold question of its cost, and whether cost increments were concealed, was very relevant to the government's announced '$465 million' total project cost.

"It is important for the reader to understand the difference between Present Value (PV) and Future Value (FV) numbers. Manipulation involving the use of these numbers was part of the government's cover-up."

The document goes on from there. Another document prepared by the South Florida Water Management District begins this way:

> PURPOSE:
> Mr. George Barley, Chairman, The Everglades Trust, requested that an independent study be made of SFWMD's approved 1994–95 Budget to determine how much of the expenditures were attributable to cleaning up pollution in the EAA. A condition of the request was that the study be fair and objective and able to withstand peer review.
>
> It was determined after an initial review of the budget documents and interviews with operating department heads that developing the expenses applicable to pollution cleanup would not be possible. This is because the budget development is not geared to segregating costs for specific requests of this nature.

That must have gotten George fuming. When I try to imagine how he might have envisioned a restored Everglades, I think of this binder, which represents his rage over the ongoing plunder of the river of grass, a vital drinking water source and symbol of our identity as Floridians. I think of these same bureaucrats and politicians, rather than working for the special interests that wield an unfair influence, working for us, the people who pay them and elect them into office. And I imagine the slow-coursing sheet flow of the river of grass, pouring unbothered into a resplendent Florida Bay, together comprising a vast watershed that is both majestic and humble, where the wildlife and all of the warring stakeholder groups—the sugar growers, environmental advocates, urban dwellers, and Miccosukee Tribe—finally have found a way to live together in peace.

In 1993 George Barley wrote to David Lawrence, then publisher of the *Miami Herald*.

"I am from the private sector," the fax began, "in real estate in Orlando, spending all of my time on matters involving the Florida Keys National Marine Sanctuary, Florida Bay (especially), and Everglades restoration."

Barley wanted to thank Lawrence for the newspaper's coverage of Florida Bay and push for more, because he felt the government agencies were not doing enough.

"Each time I go to a meeting it is like realizing you are at a movie you recently saw," George wrote.

"I am a pretty bottom line person, and to me the only measure of success in restoring a sick ecosystem is the health of the patient. Despite numerous announcements, coalitions formed, annual meetings, awards, etc. the Everglades has become sicker each year, so I think these efforts have met with wide scale failure. We need to acknowledge this and find a way to get ahead of the curve.

"My environmental friends like Charles Lee, Jim Webb, Paul Parks—heroes who have been toiling for years with inadequate thanks to restore the Everglades ecosystem—get testy when I ask them to write down a list of times the patient got better. Not when land was purchased, or money authorized, or programs announced. I am talking about net physical improvement. Well, they say, we kept it from getting worse, which they did.

"My hunting and fishing buddies talk about 'the good old days.' When were they? We used to say the 60s, then it was the 70s. At the rate we are going in 2010 our children will be saying the 90s."

20

LOOKING-GLASS WATER

The water glowed aquamarine, its ripples sparkling beneath the resplendent sunshine like a billion blinding Christmas tree lights. But the water was turbid.

This was not the looking-glass water that long had lured anglers like George Barley here for the thrill of spying a silvery tarpon gliding among the undulating seagrass, hunting the animal and then casting the line to position the bait just so to hook the fish. Then reeling in the animal: old man and the sea, Ahab and the whale.

No, the water was turbid, so much so that sight-fishing here would have been impossible, and even if George still were alive and had been able to see through the shallow water to the bottom, there was hardly any seagrass anyway. Instead the bottom was muddy, like wet clay with bits of vegetation stuck in it. With no seagrass securing the sediment it was free to swirl, cloud the water, and conceal its secrets.

Nearly thirty years after pea soup algae in this very spot of Florida Bay had derailed George's birthday fishing trip with Mary Barley and sent him into a battle that eventually ended his life, Mary and I returned to Sandy Key Basin, way in the Florida Bay backcountry. We wanted to see what all of the elapsed time, all of the legal bickering and political haggling and billions of dollars spent on restoration work, indeed the campaign George had given his life for, had meant to the situation here, this spot that had helped propel him into action. Joining us on the 21-foot skiff this sunny morning of January 2020 were two scientists who had known George and helped inform his advocacy: Jim Fourqurean and Tom Frankovich, both Florida Bay experts at Florida International University (FIU).

Sandy Key is a spit of beach and red and black mangrove trees on Florida Bay's western edge, where the bay converges with the Gulf of Mexico. Flamingo and Cape Sable are visible in the distance. As Frankovich slowed the skiff to a stop we came upon a flock of pelicans and cormorants nose-diving to the water, crashing into the surface, and feeding upon a school of small herring. For several minutes we watched the birds on their collision courses of feathers, water, and fish. Splash! Splash! Splash! After a while in the opposite direction a loggerhead sea turtle poked its nose above the surface right near the boat. We pointed. Look!

Sandy Key is situated in one of forty-seven basins in Florida Bay, each defined by chains of islands and shallow banks of seagrass, with varying levels of salinity influenced by flows from the Gulf of Mexico, Atlantic Ocean, and Everglades. In Sandy Key Basin the water is salty and, despite its turbidity, highly productive because of its proximity to the gulf, a source of phosphorus. Many guides live in Islamorada but bring their clients out here to fish.

Old photographs document clear water here, but in 1987 guides began reporting growing patches of dead seagrass just to the north. At first scientists believed the die-offs were a natural phenomenon, part of a natural cycle. Yet never before had they seen anything like it anywhere on Earth: widespread seagrass die-offs in clear water, where there were no harmful algae blooms to shade the plants from the sun, their source of life. Billions of blades floated to the surface, the muddy barren bottom visible through the looking-glass water. The cause was a mystery.

Then the harmful algae blooms did occur, nourished by nutrients released from the decaying dead seagrass. The blooms blocked even more seagrass from their life-sustaining sun, initiating a cyclical death spiral that would lead to a decade of turbid water. Eventually scientists would come to understand the spiral's origin: in the late 1980s Florida Bay was hot and salty, conditions that cause the water to lose oxygen. Without enough oxygen the seagrass died.

"How could this happen?" George had demanded that pea soup day. The answer, basically, was that Florida Bay was not getting enough freshwater from the ailing Everglades.

Today the pea soup algae was gone, and more freshwater was on the way. In the Everglades there were signs the restoration was working, despite the effort's woefully slow pace. The Kissimmee River was rebounding. Construction was under way on new reservoirs that would replenish the river of grass, and the stormwater treatment areas were cleansing the water better than anyone expected, although there were not enough of the filter marshes. The South Florida Water Management District reported some 90 percent of the water in the Everglades Protection Area was at 10 parts per billion.

Right here, though, in this fateful spot the water was turquoise but turbid. After all of this time and effort and hand-wringing, Florida Bay still was getting only 20 percent of the freshwater it once did, and there was no plan for restoring it all. Fourqurean didn't know whether it was even possible.

"I'm not sure restoration has reached here at all," said Fourqurean, a tall affable man who wore a khaki ball cap, black sunglasses, and a navy FIU polo shirt. "This basin in the past was a lush seagrass meadow, and people used to sight-fish for tarpon here and collect bay scallops. It's not that way now."

Perhaps this would have inflamed George all over again. Perhaps he would have gone into another battle, banging on politicians' doors and flying them around, hounding journalists like me. By now, though, Mary had more than twenty years of political experience on her late husband. She was the grizzled veteran. Sitting there on the bobbing skiff, dressed in a lavender long-sleeved shirt, pale-blue vest, and silver sunglasses, she seemed nonchalant, optimistic even.

"It's a work in progress," she said. "It's an opportunity to fix the plan if the plan is wrong."

There was reason for Mary's optimism.

IN 2018 SOMETHING AMAZING HAPPENED in Florida. No more were political candidates accepting campaign contributions from Big Sugar. In fact they were shunning the cash and shaming any candidate who actually took the money, now as toxic as the algae infesting Florida's waters.

In the hotly contested gubernatorial race only one candidate actually accepted the funds: state agriculture commissioner Adam Putnam. The Republican had a long history with sugar growers; he even had halted a Department of Education effort to ban chocolate milk and other sugary drinks from Florida schools and then taken over the school breakfast and lunch programs.

"I support our 'Glades communities,'" the boyish-looking red-haired Putnam proclaimed proudly from the campaign trail, adding that growers were a "vibrant part of our economy, and the water that leaves sugar farms is cleaner than the water that comes to them."

Putnam accepted millions of dollars from sugar growers before he was voted out in the August primary of the gubernatorial race, halting a more than twenty-year political career that had included terms in the state legislature and Congress.

Meanwhile all five Democratic candidates refused any contributions from the growers, although during one debate Chris King, a Central Florida businessman, called out Gwen Graham, daughter of former US senator and governor Bob Graham.

"You have taken money from the sugar industry," King accused, prompting Graham to confess she had accepted $17,000, but she said she had donated the money to an advocacy group for the Indian River Lagoon. She also suffered a breathtaking primary upset.

In the lead-up to the general election most environmental groups endorsed Democratic nominee Andrew Gillum, a former mayor of Tallahassee, but Republican nominee Ron DeSantis, a US representative, landed two important endorsements: one from President Donald Trump, who in tweets had characterized climate change as a "hoax," and Mary Barley's influential Everglades Trust, which recognized DeSantis's opposition in Congress to the sugar program.

"The Everglades and coastal estuaries couldn't care less about partisan politics, and so the trust doesn't. They are in desperate need of a hero—and they found one in Ron DeSantis," said Kimberly Mitchell, the trust's executive director. "Ron understands the critical infrastructure projects that must be undertaken and expedited, with the ability to make them a top priority, and already has a track record of standing up to an industry that is physically and politically blocking the reconnection of Lake Okeechobee to the Everglades—Big Sugar.

"Floridians have had enough of rhetoric and broken promises from our politicians. 'I will stand up to the special interests,' is what we're told in an election year. Well, we now have a politician who has actually walked the walk, and for the millions who depend on a healthy Everglades, and all the critters that call them home, it could not come soon enough."

The Everglades Trust's endorsement of DeSantis, Mary would tell me later, was based largely on his responses to a questionnaire the trust distributed to every political candidate in the state. Gillum did not fill out the questionnaire, she says.

DeSantis "ran on our issue," Mary says. "He ran on the Everglades and clean water. That was more than probably 60 percent or 70 percent of his campaign, was the environment."

DeSantis campaigned hard against sugar growers. In the last debate before the election he claimed he was "the only candidate who fought Big Sugar and lived to tell about it." In November 2018 he was elected governor. Right before the election Paul Tudor Jones contributed $250,000 to DeSantis's political action committee. Mary says DeSantis's election represented a remarkable moment in Florida political history.

"People realized they didn't have to do Sugar's bidding for them, that you could actually win and be against them and be for the Everglades," she says. "So that was just a turning point that really we needed. It was a shock, but it was great. It was fantastic."

DESANTIS TOOK OFFICE in January 2019, and two days later he unveiled a sweeping executive order aimed at toxic algae. The order called for, among other things, $2.5 billion over four years for Everglades restoration and a task force on blue-green algae. The new governor also wanted to speed up construction of the reservoir south of Lake Okeechobee that had been so controversial and hire the state's first chief resilience officer and chief science officer so "we're doing sound science." He said he expected support from Trump, who frequently visited the president's Mar-a-Lago Club in Palm Beach. Later in 2019 Trump actually would declare himself a Florida resident and announce his reelection campaign in Orlando, at the center of this crucial swing state.

DeSantis also called for resignations from every board member of the South Florida Water Management District, the state agency overseeing Everglades restoration. He said he felt the board members did not understand the toll of the toxic algae on hard-hit communities. The board members were seen as closely aligned with sugar growers.

"I just want good people who are willing to do the right thing," DeSantis said.

DeSantis replaced the board members with appointees supported by environmental groups, such as "Alligator" Ron Bergeron, a millionaire outdoorsman who had first visited the Everglades seventy-two years before, and Jacqui Thurlow-Lippisch, among the most passionate advocates I have ever encountered. Thurlow-Lippisch spent weekends flying above the watershed with her pilot husband and posted photos of the environmental fallout on her blog.

Meanwhile Agriculture Commissioner Nikki Fried, a Democrat, appointed a director of agriculture water policy and announced the department would be reviewing agricultural best-management practices. The developments represented a stunning reversal from former governor Rick Scott, who was elected to the US Senate at the same time DeSantis became governor. Scott was seen as closely aligned with sugar growers, had slashed state environmental spending, and famously banned state agencies from using words like *climate change*. Environmental groups were giddy, although some pointed out that while DeSantis now was using the words *climate change* he was doing little to address human-engineered emissions.

"It's a little bit like Christmas morning," one Audubon leader said.

Added Eric Eikenberg, chief executive officer of the Everglades Foundation, "Here is a governor who is relying on science, and as a science-based organization that is what we've been yearning for."

DeSantis followed through on his promises. He joined with other top Republican leaders in the state, including US senators Rick Scott and Marco

Rubio and US representative Brian Mast, who represented the East Coast communities hardest-hit by toxic algae, to call on Trump to spend more on Everglades restoration. The leaders prompted the president in March 2019 to tour Lake Okeechobee. During the visit Trump eyed the earthen walls of the lake's dike and said, "I'm looking at all of the walls and saying, Don't forget our southern border.'" The comment was in reference to the president's controversial plan at the time to stem illegal immigration across the country's southern border by constructing a physical barrier. In December the president signed a federal budget containing $200 million for the Everglades. By the end of the year DeSantis was Florida's most prominent advocate for environmental restoration.

Mary Barley was hardly done, though. In February 2020 she spoke out at a community meeting in Islamorada, where the US Army Corps of Engineers presented a plan that would have left Florida Bay with more freshwater during the wet season rather than in the dry season, when the bay needs it most.

Cartoon by Chan Lowe, *Sun Sentinel* editorial cartoonist in Fort Lauderdale, Florida

Courtesy of Mary Barley

"I am really appalled," she said. "We are part of Florida. . . . We started Everglades restoration. Islamorada was where it all started. . . . This is war, and the Florida Keys are going to win."

RIGHT BEFORE FLOWING INTO FLORIDA BAY the river of grass splits in two. Shark River Slough carries the water to the west, and the much smaller Taylor Slough bends to the east.

In 2012 workers completed construction on a Comprehensive Everglades Restoration Plan project aimed at improving the flow through Taylor Slough. The bureaucratically named C-111 Spreader Canal Western Project was designed to address problems caused by the C-111 canal, dug during the 1960s for the transport of rockets from an Aerojet plant in south Miami-Dade County over to Biscayne Bay and then up to Cape Canaveral. The canal upended the river of grass's natural sheet flow, rerouting the water through a drainage channel and out to Barnes Sound. The result was that the Everglades were too dry, and Florida Bay was too salty.

The C-111 Spreader Canal Western Project involved a network of pumps, levees, canals, and wetlands constructed near the Homestead entrance of Everglades National Park. Its completion coincided with a nice little spell of rain, and within a year Florida Bay began to respond. The amount of freshwater through Taylor Slough doubled, and underwater plants flourished, increasing their cover by five times. Salinities declined and the water was clearer.

"It goes to prove if we get more freshwater, conditions improve regardless of where it comes from," said Tom Frankovich, one of the Florida Bay experts who knew George and accompanied Mary and me that day in January 2020 when we returned to Florida Bay to retrace some of George's path there.

On our way out to Sandy Key Basin, Frankovich stopped the skiff in Rabbit Key Basin. I peered over the boat's side and nearly gasped.

Here was the looking-glass water.

The water was so clear, I could see straight to the bottom, which was covered with a beautiful meadow of turtle grass. It was amazing. I have been on boats all my life here in Florida, and I don't know that I have ever seen water so clear. You would not have known it was there if not for the sun's reflection on the surface.

"There is a lot of life here," Frankovich said.

The seagrass was dense, lush, its ribbon-like blades undulating with the water.

It was not a bay of grass, but it was a start.

"This looks fantastic," Mary proclaimed. "It's beautiful, really. I didn't even know there was anything left like this, seriously."

No tarpon in sight, but still George Barley would have loved it.

ACKNOWLEDGMENTS

SOME TEN YEARS AGO I PHONED Mary Barley and asked whether I could write a book about her life. It was not a simple question, and her answer of yes represented a remarkable gesture as she handed over her life, basically, to someone she hardly knew. But perhaps this was not remarkable at all for Mary, a woman who already had shared so much of herself and her life with the state of Florida, which is not even her native state.

A little while later I drove down to Islamorada to get going on the book and, after a few days, returned home with several boxes of Mary's personal records. The boxes included original glossy photographs of the Barleys smiling from far-flung spots across the globe, of George's tear-soaked memorial service—some of the happiest and most horrific moments of Mary's life, frozen in time. She had not only handed over her life, but entrusted me with these very personal details. For her singular generosity with me and the state of Florida, I am profoundly grateful.

The years following that fateful phone call were the most trying of my life professionally and personally, but Mary's story stayed with me. I mean, how could it not? Also steadfastly by my side during this difficult time was my agent, Craig Kayser, who worked relentlessly on my behalf. Without you this book would not exist.

A river of thanks to my three readers, who reviewed drafts and provided essential feedback, encouragement, and friendship during the many weekends I spent at home alone in sweats wondering whether this thing was any good: Bill Maxwell, Patti McCracken, and Neil Santaniello. I could not have done this without you. Cynthia Barnett, Trevor Aaronson, and Randy Mejeur also read early pages and offered critical help.

To re-create events that occurred some twenty and thirty years ago I relied on the memories, expertise, and personal records of many individuals who gave generously of their time, among them Paul Tudor Jones, Charles Lee, Clay Henderson, Rick Roth, Ron Jones, Curt Kiser, Fowler West, Jon Mills, Chris Ball, Lauren Barley, Kevin Barley, Jim Fourqurean, Tom Frankovich, Paul Gray, Gary Goforth, Tom Van Lent, Stephen Davis, Eric Eikenberg, Eric Draper, Ramon Iglesias, Sean Cooley, Randy Smith, John Campbell, Nathaniel Reed, and Hank Brown. I also am grateful to the farmers and residents of the Everglades Agricultural Area who took the time to talk with me even as they asked that their names not be included. I have worked to give them a voice within these pages while honoring their wishes.

Special thanks to WMFE (NPR Orlando), a radio station that somehow found a place in its newsroom for a battered and weary print journalist. Together we are a scrappy team of journalists covering the day's news with diligence and good humor. I especially thank Matthew Peddie, Nicole Darden Creston, Brendan Byrne, Abe Aboraya, Danielle Prieur, Talia Blake, Joe Byrnes, Crystal Chavez, Renata Sago, Emily Lang, Mark Simpson, Judith Smelser, and Tom Parkinson. I have learned from each of you, and this book is better for it.

How fortunate I am to have as my publisher Johns Hopkins University Press and my editor, Tiffany Gasbarrini, one of those rare individuals who is smart, professional, and kind all rolled into one. Thank you for advocating for my work, always. Thanks also to Steven Baker for his careful copyediting, Esther Rodriguez for her attention to detail, Kathy Patterson for her index work, and Juliana McCarthy and the entire Johns Hopkins team for working with me on production deadlines during a historic pandemic.

I am grateful for the support of the Florida Center for Investigative Reporting, Burrow Press, United Arts of Central Florida, the *Christian Science Monitor*, and *Newsweek*.

For her unbounding encouragement and friendship I thank Kristin Harmel, who helped me navigate this crazy world of book publishing! For their friendship and support I also thank Michael Schneider, Tristram Korten, Beth Kassab, Nancy Beyer, Rachelle Mirizio, Vanessa Loomie, and Besa Kosova and her dream that fights.

My very dearest thanks go to my family, who provided endless assistance, encouragement, and affection: Ron and Cathy Burkdoll and Kristy, Mark, Gavin, and Teagan Bower. I treasure you all. And most profoundly Rebecca: everything I do is for you, sweet girl.

TIME LINE

1948 The US Congress authorizes the Central and Southern Florida Project for Flood Control and Other Purposes, or C&SF Project. The effort is hailed at the time as the nation's largest civil works project and, along with earlier projects, drains the Everglades to half its historic size.

1987 A seagrass die-off and harmful algae bloom mark the start of ongoing widespread problems in Florida Bay, a collapse one scientist describes as "unprecedented in history."

1988 The federal government sues the state of Florida. Specifically, the lawsuit alleges the South Florida Water Management District and state Department of Environmental Regulation (as it was called at the time) are violating their own water quality laws by allowing sugar growers in the Everglades Agricultural Area to pollute Everglades National Park and the Arthur R. Marshall Loxahatchee National Wildlife Refuge.

1991 Governor Lawton Chiles "surrenders his sword." The federal government and state reach a settlement, and seven months later a federal judge signs a thirty-six-page consent decree ordering the state and sugar growers to implement a cleanup plan.

1992 The Florida Keys National Marine Sanctuary Advisory Council is formed, with George Barley as its chair.

1992 With South Florida's booming population exerting growing pressure on the fragile Everglades, Congress calls for a "restudy" of the C&SF Project as part of the Water Resources Development Act.

1993 An Everglades cleanup plan is introduced. Sugar magnate Alfonso Fanjul declares, "Today the Clinton administration delivers," but environmental groups blast the plan and charge that the administration has not delivered anything substantive.

1993 George Barley launches penny-a-pound. The following May the Florida state supreme court decides the referendum language, among other things, smacks of "political rhetoric" and declares the measure unconstitutional.

1994 State lawmakers approve the Everglades Forever Act. Environmental groups denounce the measure; it is renamed from the "Marjory Stoneman Douglas Act," after the author-turned-activist wrote to the governor to demand that her name be removed from its title. "I disapprove of it whole-heartedly," she writes of the legislation.

1995 While on his way to meet with the US Army Corps of Engineers about the Everglades, George Barley dies in a plane crash. A bereaved Mary Barley jumps into his advocacy work.

1996 Penny-a-pound becomes the most expensive political campaign at the time in Florida history. Voters reject penny-a-pound but approve two related state constitutional amendments: one requiring polluters to pay for their own cleanup and another establishing a trust fund.

2000 President Bill Clinton signs the Comprehensive Everglades Restoration Plan (CERP) into law. The effort is billed as the world's largest restoration effort ever.

2002 Mary Barley switches her party affiliation from Republican to Democrat and runs for commissioner of agriculture and consumer services. Agricultural groups line up against her, and she loses in the primary to a schoolteacher-librarian.

2007 Construction begins on the first CERP project, the Picayune Strand. Meanwhile the US Government Accountability Office projects the restoration effort's cost at $10 billion, up from $7.8 billion, and suggests that additional efforts, such as restoration of the Kissimmee River, could push the cost of restoring the Everglades to as much as $20 billion.

2008 Governor Charlie Crist announces the state has agreed to a $1.75 billion deal to buy out U.S. Sugar Corp. and put the land toward Everglades restoration. Two years later, with the worst recession in a generation deepening, the South Florida Water Management District closes on 26,800 acres for $197 million.

2016 Toxic algae prompts a proposal for a reservoir south of Lake Okeechobee for Everglades restoration. Sugar growers oppose the reservoir. Eventually lawmakers approve the reservoir to be built on state land.

2018 The National Academies of Sciences, Engineering, and Medicine calls for a "mid-course assessment" of CERP to account for climate change, an issue the scientists say has heretofore been neglected in the planning. The scientists point out the effort's slow pace, putting the end date well beyond 2080.

2018 Governor Ron DeSantis is elected after campaigning against toxic algae and Big Sugar. Two days after taking office he unveils a sweeping executive order that, among other things, calls for $2.5 billion to be spent over four years for Everglades restoration. Later, DeSantis demands resignations from the entire board of the South Florida Water Management District. The board members were seen as closely aligned with sugar growers.

2019 One CERP project is complete, components of multiple projects are finished, and three projects are in planning. Everglades restoration remains one of the world's most substantial efforts at ecological restoration, and CERP is now projected to cost up to $17 billion.

NOTE ON SOURCES

FORMING THE BACKBONE OF MY RESEARCH material for this book were news articles spanning the 1980s to the present, and I owe a debt of gratitude to Florida's major newspapers, which did such a thorough job of covering these extremely complex issues, especially the *Miami Herald*, *Orlando Sentinel*, *Tampa Bay Times*, *Fort Lauderdale Sun Sentinel*, and the *Palm Beach Post*.

Any quoted material represents words said to me, words as they appeared in personal records or official documents, or words as they were quoted in news articles. I've also quoted some dialogue as it was recounted to me during interviews as sources described past events. Wherever possible I conducted multiple interviews to ensure accuracy.

Introduction

Much of the first section of the introduction is based on an interview with Tom Van Lent and also my personal experience in the Everglades, namely in Big Cypress National Preserve and Everglades National Park. For the anecdote about the Old Ingraham Highway I relied on *Gladesmen: Gator*

Hunters, Moonshiners, and Skiffers, by Glen Simmons and Laura Ogden. I also read and enjoyed "Evolve: A Case for Modernization as the Road to Salvation," by Michael Shellenberger and Ted Nordhaus, which appeared in *Orion* magazine in October 2011. *The Swamp: The Everglades, Florida, and the Politics of Paradise*, by Michael Grunwald, provided important context. The second section is based on interviews with Mary Barley for *Newsweek* and the *Christian Science Monitor*.

CHAPTER 1. *George Barley's Birthday*

I wrote my description of Florida Bay and Islamorada after a media tour of the bay hosted by the Everglades Foundation and also several trips to see Mary at her home. The anecdote about George Barley's birthday comes from interviews with Mary and Hank Brown. Much of the description of Florida Bay's decline comes from George's own writings while he was chairman of the Florida Keys National Marine Sanctuary Advisory Council. Also helpful were "The Last Days of Florida Bay," by Carl Hiaasen, which appeared in *Sports Illustrated* magazine on September 18, 1995, and "The Dead Zone," by Dan Sewell of the Associated Press, which appeared in the *Miami Herald* on April 18, 1993.

CHAPTER 2. *The Big Picture*

My summary of Everglades history is based on a number of sources, especially *The Everglades: River of Grass*, by Marjory Stoneman Douglas; *The Everglades Handbook: Understanding the Ecosystem*, by Thomas E. Lodge; and *The Swamp: The Everglades, Florida, and the Politics of Paradise*, by Michael Grunwald. To reconstruct George Barley's advocacy during the early 1990s I relied on his own writings and also my interviews with Ron Jones. For further information on Florida Bay's decline I referred to the letters George Barley asked Ron Jones and other prominent scientists to write. All of these documents were part of Mary Barley's personal records. Also helpful was "Saving Florida Bay," by Heather Dewar, which appeared in the *Miami Herald* on August 15, 1992.

CHAPTER 3. *Big Sugar*

The anecdote about the plane trip comes from an interview with Ron Jones. The history of sugar in Florida and description of the Everglades Agricultural Area come from a number of sources, especially the *Florida Sugarcane Handbook*, a publication of the University of Florida; U.S. Sugar Corp. fact sheets; and interviews with Rick Roth. Also helpful was "In the Kingdom of Big Sugar," by Marie Brennar, which appeared in *Vanity Fair* in February 2011; "Big Sugar," by Karl Vick, which appeared in the *St. Petersburg Times* (now the *Tampa Bay Times*) on May 15, 1994; and "Sugar Field Burning Plagues Poor Florida Towns with Soot," by Ellis Rua of the Associated Press, which appeared on WUSF.org on December 1, 2019. To reconstruct the very complicated chronology of the federal government's litigation against the state of Florida over sugar growers' pollution in the Everglades, I relied on many sources. Especially helpful were multiple articles by Heather Dewar of the *Miami Herald*. Also helpful was "Sugar Industry: Lehtinen Trying to Run Us Out," by Mary McLachlin, which appeared in the *Palm Beach Post* on October 10, 1989. A *Palm Beach Post* op-ed entitled "Don't Let Technicalities Kill Everglades Cleanup," published on September 3, 1995, provided a very helpful chronology. George Barley's letters to Dexter Lehtinen are part of Mary Barley's personal records. My description of the Clinton administration's unveiling of a $465 million plan aimed at saving the Everglades comes from multiple sources, including "Agreement Would Clean Up the Everglades," by Karl Vick, which appeared in the *St. Petersburg Times*. Michael Grunwald's *The Swamp: The Everglades, Florida, and the Politics of Paradise* provided important context, as did multiple farmers and residents of the Everglades Agricultural Area who asked that their names not be included in this book.

CHAPTER 4. *The Politics of Water*

I wrote the opening scene of Lake Okeechobee after a boat tour with Ramon Iglesias. Background information on toxic algae in the lake and elsewhere is based on my own reporting for WMFE (NPR Orlando). The summary of how the US Army Corps of Engineers and South Florida Water Management District manage the Everglades' waters is based on multiple interviews, especially with John Campbell of the army corps and Paul Gray of Audubon Florida. I also referred to the Lake Okeechobee Regulation Schedule. I interviewed Garth Redfield to learn more about how the Everglades' water management compares with that of other watersheds across the globe. *Rain: A Natural and Cultural History*, by Cynthia Barnett, provided important context, as did other sources. I also read and enjoyed "Water Maze: Inside the Circulatory System of a Desert City," by Craig Childs, which appeared in *Orion* magazine in October 2015. Also helpful was the 2018 *South Florida Environmental Report*, a publication of the South Florida Water Management District. I dialed into one of the US Army Corps of Engineers' periodic scientific calls to hear some of the conversation behind the corps's water management decisions. Further background on Lake Okeechobee and toxic algae comes from "Understanding the Effect of Salinity Tolerance on Cyanobacteria Associated with a Harmful Algal Bloom in Lake Okeechobee, Fla.," a scientific report of the US Department of the Interior and US Geological Survey. Also helpful was "Political Muscle Halts Release of Lake O's Foul Water: The Reprieve Won't Last Long," by Jenny Staletovich, which appeared in the *Miami Herald* on July 9, 2018. The chapter's last section is based on my interview with Mary Barley. For background on Florida Bay's condition I interviewed Stephen Davis of the Everglades Foundation.

CHAPTER 5. *The Campaign Begins*

The point about Florida's waters being held in a public trust comes from the op-ed "Florida's Water Laws Have Not Been Enforced," by Robert Knight, which appeared in the *Gainesville Sun* on June 20, 2018. I interviewed Eric Draper to learn about the problems confronting the Everglades in 1993 and also about George Barley's thinking at the time. The scene of George Barley talking with Mary about his political plans comes from interviews with Mary. I relied on many sources to reconstruct George's penny-a-pound campaign, including "Tax Proposal Could Stall

Everglades Talks," by Carolyn Fretz and Mary Ellen Klas, which appeared in the *Palm Beach Post* in September 30, 1993; "Big Sugar Is In for a Big Fight," by Jon East, which appeared in the *St. Petersburg Times* (now the *Tampa Bay Times*) on October 3, 1993; and "Growers Unleash Lawyers to Stop Penny Sugar Tax," by Kirk Brown, which appeared in the *Palm Beach Post* on May 2, 1994. I also referred to George's own writings. His 1993 memo was part of Mary's personal records. The anecdote about Vee Platt comes from "A Way of Life," by Jeff Klinkenberg, which appeared in the *St. Petersburg Times* on May 16, 1994. Charles Lee provided important context about penny-a-pound. The summary of the Everglades Forever Act is based on multiple sources, including *The Swamp: The Everglades, Florida, and the Politics of Paradise*, by Michael Grunwald, and "Gov. Chiles Signs Everglades Bill as Debate Rages," by William Booth of the *Washington Post*, which appeared in the *Chicago Sun-Times* on May 4, 1994. Reactions inside the penny-a-pound campaign to the state supreme court's ruling against the ballot initiative come from interviews with Jon Mills and Mary Barley. I also referred to the campaign's news release at the time, which was part of Mary's personal records. Also helpful was "Court Bars Glades Sugar Tax from Ballot," by Terry Neal, which appeared in the *Miami Herald* on May 27, 1994.

CHAPTER 6. *The Barleys*

The opening scene of Chris Ball's job interview comes from interviews with Ball and Mary Barley. George Barley's biographical information comes from multiple sources, including interviews with Mary, his daughter Lauren Barley, and his nephew Kevin Barley, a family historian. Also helpful were "Loss of Fishing Makes Developer Environmentalist," by Pat Beall, which appeared in the *Orlando Business Journal* on November 5, 1993; "Barley Guards over Everglades—Developer Plays Leading Role in Trying to Save Sensitive Land," by Craig Quintana and Kevin Spear, which appeared in the *Orlando Sentinel* on April 18, 1994; "Orlando Realtor Steps Up to Battle Big Sugar," by Elizabeth Willson, which appeared in the *St. Petersburg Times* (now the *Tampa Bay Times*) on March 23, 1994;

"Crusading Developer: Healthy Environment Key to Prosperity," by Carolyn Fretz, which appeared in the *Palm Beach Post* on October 11, 1993; and "Developer Who Fought 'Holy War' to Save Everglades Dies in Crash," by Tom Fiedler, which appeared in the *Miami Herald* on June 24, 1995. For information on George Barley's father I relied on "George Barley Sr., Former Mayor, Dies," which appeared in the *Orlando Sentinel* on April 11, 1992. For information on George's early political affiliations I referred to "15 Named to Sanctuary Committee," by Dan Keating, which appeared in the *Miami Herald* on February 7, 1992; "Barley Jr.: Political Pressures Cause Delay," by Don Wilson, which appeared in the *Orlando Sentinel* on April 13, 1986; and "Sanctuary Opponents Unleash Ire," by Dan Keating, which appeared in the *Miami Herald* on July 24, 1992. Mary Barley's biographical information comes from multiple sources, including interviews with Mary and her brother Dave Wilmot and also Mary's writings, which were included in her personal records.

CHAPTER 7. *The Fanjuls*

The anecdote about Monica Lewinsky and President Bill Clinton comes from special prosecutor Kenneth Starr's report to Congress. The Fanjul brothers' biographical information comes from multiple sources. Especially helpful were "In the Kingdom of Big Sugar," by Marie Brennar, which appeared in *Vanity Fair* in February 2011, and "Kingdom of Cane—The Influential Fanjul Family, Made Wealthy by Sugar, Stands to Lose a Lot If the Everglades Cleanup Tax Passes," by Robert McClure and David Beard, which appeared in the *Fort Lauderdale Sun Sentinel* on October 20, 1996. I also referred to "Fanjuls Often Criticized for Treatment of Workers," by Jim McNair, which appeared in the *Miami Herald* on October 7, 1991; "Often in the Public Spotlight, Fanjuls Put Premium on Privacy; Sugar-Production Is Closely Watched," by Rick Christie, which appeared in the *Miami Herald* on January 27, 1991; "Sugar Growers Reap Bonanza in Glades," by Sean Holton, which appeared in the *Orlando Sentinel* on September 18, 1990; "Report Reveals Fanjul's Influence," by Robert P. King, which appeared in the *Palm Beach Post* on September 13, 1998; "First Family of Florida Sugar," which appeared in the *Palm Beach Post* on December 19, 1999; and "The People Who Comprise Big Sugar," which appeared in the *Fort Lauderdale Sun Sentinel* on April 25, 2004.

CHAPTER 8. *Big Special Interests*

The opening scene comes from the *60 Minutes* story on the sugar program. To explain how the sugar program works and its economic impact, I relied on multiple sources, including my own exhaustive reporting for my article "In Sugar Price Supports, Sour Tastes for Consumers," which the Florida Center for Investigative Reporting published on September 9, 2012. Another significant source was "The Sweet Smell of Excess," by John Dorschner, which appeared in the *Miami Herald* on January 28, 1990. Also helpful were "Sugar Growers Reap Bonanza in Glades," by Sean Holton, which appeared in the *Orlando Sentinel* on September 18, 1990; "Big Sugar Sweetens Candidates' Coffers—Sugar Funds—The Sugar Program," by William E. Gibson, which appeared in the *Orlando Sentinel* on June 20, 1993; "U.S. May Tie Sugar Loans to Cleanup," by Kirk Brown and Lisa Shuchman, which appeared in the *Palm Beach Post* on March 11, 1994; "Sweet Deal for Sugar Industry Going Sour," by Lisa Shuchman, which appeared in the *Palm Beach Post* on April 9, 1995; "Sugar's Free Ride May Not Be as Sweet: A Reform-Minded Congress This Year Might Go after the Federally Controlled Program," by Sean Holton, which appeared in the *Orlando Sentinel* on May 14, 1995; and "Growers' Sweet Life May Come to an End: Sugar Program Faces Challenge," by Tom Fiedler, which appeared in the *Miami Herald* on May 21, 1995. The anecdote about Charles Lee's tour of Florida newspaper offices comes from interviews with Lee and also his personal records. I also referred to Mary Barley's personal records. I interviewed Michael Wohlgenant and Lance deHaven-Smith for the Florida Center for Investigative Reporting. Further information came from interviews with Rick Roth. The scene of the congressional subcommittee meeting comes from interviews with Fowler West and also from "Sugar Towns Say Way of Life Threatened," by Tom Fiedler, which appeared in the *Miami Herald* on April 28, 1995.

CHAPTER 9. *The Plane Crash*

The anecdote about George Barley missing his flight at the Orlando International Airport came from interviews with Mary Barley. Details of the crash itself came from official reports from the Orange County Sheriff's Office and National Transportation Safety Board. Information about the subject of George's meeting with the US Army Corps of Engineers came from interviews with Lewis Hornung and Charles Lee. Background information on the Everglades restudy came from "Glades Plan Making Waves: Farmers, Fishermen Are at Odds over Cleanup," by Heather Dewar and Lori Rozsa, which appeared in the *Miami Herald* on December 19, 2003; and "Corps Takes On a New Mission: Rescue 'Glades—'River Of Grass,'" by Robert McClure, which appeared in the *Fort Lauderdale Sun Sentinel*, on May 4, 1993. The anecdote about how Mary learned of her husband's death came from interviews with her. I also interviewed Bob Showalter, president at the time of George's death of Showalter Flying Service. Further information about the crash came from an interview with Jeffrey Kennedy of the National Transportation Safety Board. To learn about how George's family reacted to his death I talked with his daughter Lauren Barley and nephew Kevin Barley. The anecdote about the Barleys on the scuba diving trip comes from Mary Barley. The anecdote about George's voice mail on Charles Lee's answering machine comes from an article Lee wrote for *The Florida Naturalist* published in the fall of 1995. To learn about pilot Mark Swade I referred to "Heroic Pilot Had Passion for Flying—Mark Swade Guided His Crashing Airplane Away from a Day-Care Center and an Apartment Building Friday," by Susan Jacobson, which appeared in the *Orlando Sentinel* on June 25, 1995. My dad, Ron Burkdoll, a pilot for fifty years, also provided helpful insight.

CHAPTER 10. *Toleration and Process*

Mary Barley talked to me about George's death during interviews. Especially helpful as I worked to reconstruct George's stature in the Everglades movement at the time of his death, as well as his memorial service, was "Hundreds Attend Service for Barley," by Kevin Spear, which appeared in the *Orlando Sentinel* on June 28, 1995. I also referred to photographs of the memorial service and transcripts of the eulogies, which were part of Mary's personal records. Also helpful were "Barley Mixed Love for Land with Business," by Kevin Spear, which appeared in the *Orlando Sentinel* on June 24, 1995; "Developer Who Fought 'Holy War' to Save Everglades Killed in Crash," by Tom Fiedler, which appeared in the *Miami Herald* on June 24, 1995; "Glades Advocate Dies in Crash," by Bill Moss, which appeared in the *St. Petersburg Times* (now the *Tampa Bay Times*) on June 24, 1995; "Plane Crash Blame: Pilot, Equipment," by Christopher Quinn, which appeared in the *Orlando Sentinel* on November 21, 1995; and "He Cared for Earth—And Acted on It," by Diane Hirth, which appeared in the *Fort Lauderdale Sun Sentinel* on July 9, 1995. The letters were part of Mary's personal records, including the one from David Weiman. Nathaniel Reed talked to me about George Barley during an interview.

CHAPTER 11. *The Campaign Resumes*

The opening scene came from interviews with Clay Henderson and Mary Barley. To reconstruct Mary's penny-a-pound advocacy I relied on interviews with her and also news articles from the time, including "Bush to Fish Glades Tourney," by Susan Cocking, which appeared in the *Miami Herald* on August 22, 1995; "Bush Finds No Luck in Tournament," which appeared in the *Miami Herald* on August 24, 1995; "Group: Sugar Industry's Clout Robs Taxpayers," by Cyril Zaneski, which appeared in the *Miami Herald* on October 6, 1995; "Both Sides Take 'Glades Fight to Newspaper Ads," by Robert McClure and Neil Santaniello, which appeared in the *Fort Lauderdale Sun Sentinel* on October 25, 1995; "The Battle over Sugar Is Turning Bitter: Foes Joust in Media, Congress," by Tom Fiedler, which appeared in the *Miami Herald* on October 26, 1995; "Sugar Messages Slick to Sponsor, Offensive to Foe," by Lisa Shuchman, which appeared in the *Palm Beach Post* on November 9, 1995; "Sugar-Tax Ads Put Squeeze on GOP Poll Contenders—Ecology Advocates Press Republican Hopefuls on Taxing Sugar Industry," by Craig Quintana, which appeared in the *Orlando Sentinel* on November 17, 1995; and especially "Another Barley Champions Everglades—George Barley's Widow Took Over His Campaign to Clean Up Pollution," by Kevin Spear, which appeared in the *Orlando Sentinel* on December 31, 1995. I also referred to "Environmentalists Seek Sugar Tax—The 2-Cent-Per-Pound Charge Would Be Used to Restore the Everglades," by Kevin Metz, which appeared in the *Tampa Tribune* (now defunct) on October 26, 1995; "Governor, 2 Senators Support Sugar Fee," by Phil Willon, which appeared in the *Tampa Tribune* on November 4, 1995; "Sugar Farmers Protest Tax Plan: Vice President to Endorse Today," by Robin Benedict, which appeared in the *Orlando Sentinel* on February 19, 1996; "Renewing the 'Glades—Gore Plan Buoys Environmentalists, Angers Sugar Growers," by Robert McClure, which appeared in the *Fort Lauderdale Sun Sentinel* on February 20, 1996; "Gore Calls for Sugar Tax to Help Everglades," by David Dahl, which appeared in the *St. Petersburg Times* (now the *Tampa Bay Times*) on February 20, 1996; "Everglades Proposal May Be High and Dry—Congress Is in No Mood to Spend More than $1 Billion on President Clinton's Everglades Plan, Inside Sources Say," by Sean Holton, which appeared in the *Orlando Sentinel* on February 22, 1996; "Everglades Has Problems Even $200 Million Can't Solve—Money for Land Buying and Cleanups Is Flowing This Election Year, but Billions Are Needed," by Sean Holton, which appeared in the *Orlando Sentinel* on April 7, 1996; "Sugar Tax Advocates Seek Spot on Ballot," by William Howard, which appeared in the *Palm Beach Post* on April 21, 1996; "Round 2 of Sugar-Tax Fight Opens Today," by Larry Kaplow, which appeared in the *Palm Beach Post* on August 29, 1996; "Should State Tax Sugar? You'll Decide—Supreme Court Says Penny-a-Pound Questions Can Go on Ballot," by Larry Kaplow and Lisa Shuchman, which appeared in the *Palm Beach Post* on September 25, 1996; and "War over Sugar Tax Costliest in History," by Larry Kaplow, which appeared in the *Palm Beach Post* on October 22, 1996. The memo Fowler West sent to Mary Barley was part of Mary's personal records, as was the transcript of her speech at a wetlands seminar at the Department of State in Washington, DC. To learn about the strategy behind the penny-a-pound campaign I talked with Mary Barley, Fowler West, Clay Henderson, and Charles Lee. Rick Roth provided important insight into how the ballot initiative affected sugar growers and also the growers' campaign strategy. I also referred to "Sugar Tax Fight: It's Nasty and Expensive—Both Sides Offering Muck, Half-Truths," by Neil Santaniello, which appeared in the *Fort Lauderdale Sun Sentinel* on October 22, 1996; Take Sugar-Jobs Claims with a Grain of Salt, Experts Say," by Cyril T. Zaneski, which appeared in the *Miami Herald* on November 4, 1996; "Man Backing 'Glades Cleanup Accused of Polluting Waters," by Neil Santaniello, which appeared in the *Fort Lauderdale Sun Sentinel* on October 11, 1996; "Cleansing the River of Grass under New Law: U.S., Florida Split Tab," by Carol Rosenberg, which appeared in the *Miami Herald* on October 13, 1996; "Sugar-Tax Campaign Leaves Bad Taste in Voters' Mouths," by Lisa Shuchman, which appeared in the *Palm Beach Post* on November 3, 1996; "Tax on Sugar Growers Is Rejected," by Cyril T. Zaneski, which appeared in the *Miami Herald* on November 6, 1996; and "Sugar's All-Out Attack Reversed Tax's Support," by Lisa Shuchman, which appeared in the *Palm Beach Post* on November 7, 1996. The letter Vice President Al Gore sent to Mary Barley was part of her personal records. For further insight into Election Day I relied on interviews with Mary Barley, Charles Lee, and Rick Roth.

CHAPTER 12. *The Financier*

Paul Tudor Jones's biographical information came from news articles, especially "Quotron Man: Paul Tudor Jones II Swaggers and Profits through Futures Pits—He Often Moves Markets by Speculating Wildly; But Is He Too Confident?—Full Moon Theory of Trading," by Scott McMurray, which appeared in the *Wall Street Journal* on May 10, 1988; "Tycoon Brings Big Money to Sugar-Tax Campaign," by Kirk Brown, which appeared in the *Palm Beach Post* on October 27, 1993; "He Has the Money to Make His Voice Heard," by David Dahl, which appeared in the *St. Petersburg Times* (now *Tampa Bay Times*) on November 13, 1995; 'Glades Crusader Has Heart, Cash," by Neil Santaniello, which appeared in the *Fort Lauderdale Sun Sentinel* on September 22, 1996; "Millionaire Bets on

Everglades' Future," by Lisa Shuchman, which appeared in the *Palm Beach Post* on September 30, 1996; "Soured on Big Sugar, Broker Boosts Glades," by Cyril T. Zaneski, which appeared in the *Miami Herald* on September 22, 1996; and "The Savior in Question," by David Olinger, which appeared in the *St. Petersburg Times* on October 19, 1996. The information about Jones's childhood, including the anecdote about reeling in a barracuda at the Cheeca Lodge, came from interviews with Jones. I also talked with him about his friendship with George Barley and George's death. The information about the destroyed wetlands at Tudor Farms came from news articles, including "$2 Million for Ruining Wetlands: Millionaire Investor Also Banned from Hunting," by the Associated Press, which appeared in the *Orlando Sentinel* on Mary 27, 1990; and "To Get the Public's Attention, Polluters May Go to Prison," by Gil Klein, which appeared in the *Tampa Tribune* (now defunct) on April 9, 1991. Also helpful was "Tudor Settles SEC Case over Trades That Caused Industrial Average to Fall," by Laura Jereski, which appeared in the *Wall Street Journal* on September 13, 1996. Charles Lee provided important context. I also talked with Mary Barley about her friendship with Paul Tudor Jones.

CHAPTER 13. *A Big Law*

The opening scene of the test marshes came from "Betting Millions on Marshes," by Robert P. King, which appeared in the *Palm Beach Post* on December 31, 1996. Background information on the restudy came from *The Everglades Handbook: Understanding the Ecosystem*, by Thomas E. Lodge. To write about the state supreme court hearing and ruling on penny-a-pound I referred to interviews with Mary Barley and also "Pollution Question Goes to Justices—The Governor Has Asked the High Court to Define Who Is Liable for the Everglades Cleanup," by Linda Kleindienst, which appeared in the *Orlando Sentinel* on May 6, 1997; and "Court Holds Polluters 100% Liable—State Cannot Enforce Ruling; Initiative in Legislators' Laps," by Linda Kleindienst and Neil Santaniello, which appeared in the *Fort Lauderdale Sun Sentinel* on November 27, 1997. To write about the Cape Sable seaside sparrow I referred to "A Watchful Eye on the Sparrow," by Craig Pittman, which appeared in the *St. Petersburg Times* (now the *Tampa Bay Times*) on

June 3, 1999; "Threatened Sparrow's Population Soars," by Cyril T. Zaneski, which appeared in the *Miami Herald* on May 27, 1998; and "Tribe Upset over Mistake in Bird Count," by Cyril T. Zaneski, which appeared in the *Miami Herald* on August 31, 1998. To reconstruct the very complicated chronology of the restudy I relied on many sources, including "Plan to Replenish Everglades Is Bold, Risky—And Expensive," by Cyril T. Zaneski, which appeared in the *Miami Herald* on July 5, 1998; "A New Life for the Everglades—$7.8 Billion Restoration Plan Unveiled," by Neil Santaniello, which appeared in the *Fort Lauderdale Sun Sentinel* on October 11, 1998; "Everglades: What Cost Restoration?—First of Four Public Hearings on the Plan Scheduled for Tonight in Clewiston," by Robert P. King, which appeared in the *Palm Beach Post* on November 2, 1998; "Is Everglades' Rescue Realistic?" by Craig Pittman, which appeared in the *St. Petersburg Times* (now the *Tampa Bay Times*) on October 7, 1998; "Water Supply, Environment and Flood Control at Top of New Plan," by Craig Pittman, which appeared in the *St. Petersburg Times* on October 13, 1998; "Everglades Plan Faces Resistance, Suspicion," by Craig Pittman, which appeared in the *St. Petersburg Times* on November 23, 1998; and "Paying for Everglades Restoration Is Hard Part," by Craig Pittman, which appeared in the *St. Petersburg Times* on July 30, 1999. For insight into some of the behind-the-scenes discussions about the restudy and emerging Comprehensive Everglades Restoration Plan, I talked with Curt Kiser. Also helpful were "Bush Details $2.3 Billion," by Robert P. King, which appeared in the *Palm Beach Post* on January 19, 2000; "River of Money Set to Flow—Everglades Cleanup Would Get $1.25 Billion over 10 Years," by David Mark of the Associated Press, which appeared in the *Orlando Sentinel* on January 19, 2000; and "Bush Details Everglades Finance Plan," by Craig Pittman and Julie Hauserman, which appeared in the *St. Petersburg Times* on January 19, 2000. The scene of President Bill Clinton signing the Comprehensive Everglades Restoration Plan came from interviews with Mary Barley. I also referred to "2000 a Turning Point for Everglades—Environmentalists Say Battles Still Must Be Waged to Undo the Harm of Canals Built in the Past, Though," by Karin Meadows of the Associated Press, which appeared in the *Stuart (FL) News* on December 26, 2000. *The Swamp: The Everglades, Florida, and the Politics of Paradise*, by Michael Grunwald, provided important context.

CHAPTER 14. *The Politics of Science*

A very significant source for this chapter was "Dangerous Science," by John Dorschner, which appeared in the *Miami Herald* on April 28, 1996. I also referred to interviews with Ron Jones. Also helpful were "Everglades Deadline Near: Scientists Test Technology for Cleanup," by Cyril T. Zaneski, which appeared in the *Miami Herald* on February 27, 2000; "State Backs Low Level for Everglades Phosphorus," by Robert P. King, which appeared in the *Palm Beach Post* on December 8, 2001; "Choking the Glades," by Craig Pittman, which appeared in the *St. Petersburg Times* (now the *Tampa Bay Times*) on December 1, 2002; "Big Sugar vs. the Everglades," by Daniel Glick, which appeared in *Rolling Stone* magazine on May 2, 1996; and "Eye on the Everglades," by Forrest Norman, which appeared in the *Miami New Times* on January 22, 2004. The point about biodiversity came from "Why Manage Exotic Vegetation?" an article that appeared on the National Park Service website (updated July 17, 2015, https://www.nps.gov/ever/learn/nature/whymanageplants.htm). To learn about the beginnings of the Everglades Foundation I talked with Bill Riley, Paul Tudor Jones, and Mary Barley. The anecdote about the highway billboard came from "Billboards, Phone Calls: Water Fight Boiling Over," by Curtis Morgan, which appeared in the *Miami Herald* on April 26, 2003; and "Billboard Plasters Water Manager," by Robert P. King, which appeared in the *Palm Beach Post* on April 26, 2003. For insight into the 10-parts-per-billion debate during the 2000s, I referred to "White House Enters Glades Dispute," by Lesley Clark, which appeared in the *Miami Herald* on April 26, 2003; "Phosphorus Threat at Center of Debate on Glades Renewal," by Curtis Morgan, which appeared in the *Miami Herald* on May 12, 2003; "Criticized Glades Bill Signed by Bush," by Lesley Clark and Curtis Morgan, which appeared in the *Miami Herald* on May 21, 2003; Glades Fight Far from Over despite Revised Water Law," by Curtis Morgan, which appeared in the *Miami Herald* on July 21, 2003; "Judge: Glades Cleanup Ignored—A State Law That Pushed Back Deadlines for Cleaning Up the Everglades Violated the Clean Water Act, a Federal Judge Has Found," by Scott Hiaasen and Evan Benn, which appeared in the *Miami*

Herald on July 30, 2008; and "Everglades Restoration Bill Signed by Scott in West Palm," by Christine Stapleton, which appeared in the *Palm Beach Post* on May 29, 2013. To learn about how the water is monitored under the 10-parts-per-billion standard I talked with Sean Cooley at the South Florida Water Management District and Gary Goforth, an independent consultant. I also referred to "Two Farms Still Pollute Everglades—Clean-Up Rules Allow Properties on State Land to Fail Overall Goal," by Andy Reid, which appeared in the *Fort Lauderdale Sun Sentinel* on December 19, 2016. Also helpful were "Florida Has So Far Failed to Fix the Everglades, Feds Say in Legal Fight over Water," by Jenny Staletovich, which appeared in the *Miami Herald* on January 17, 2019; "Federal Oversight of Everglades Still Needed, Groups Say—Water District Board Voted in November to Seek End to Order," by Kimberly Miller, which appeared in the *Palm Beach Post* on January 20, 2019; and "Federal Judge Rejects Move by South Florida Water Managers to End Everglades Oversight," by Jenny Staletovich, which appeared in the *Miami Herald* on February 11, 2019. The anecdote about the George Barley Water Prize came from sources at the Everglades Foundation and Mary Barley. I also referred to "Solution Sought for Pollution—$10 Million Prize Offered for Fix to Water Woes," by Andy Reid, which appeared in the *Fort Lauderdale Sun Sentinel* on July 22, 2016; and "Figure Out How to Cheaply Fix Algae Blooms and Win $10 Million," by Greg Allen, which aired on National Public Radio's *All Things Considered* on July 6, 2018 (https://www.npr.org/2018/07/06/626601088/figure-out-how-to-cheaply-fix-algae-blooms-and-win-10-million).

CHAPTER 15. *Running for Office*

To reconstruct Mary Barley's campaign for agriculture commissioner I referred to interviews with Mary and also "Cabinet Hopeful Nettles Agribusiness," by Adam C. Smith, which appeared in the *St. Petersburg Times* (now the *Tampa Bay Times*) on August 5, 2002; "Barley Rocks a Quiet Contest," by Lloyd Dunkelberger, which appeared in the *Sarasota Herald-Tribune* on August 26, 2002; "Race for Agriculture Seat Livens Up at 11th Hour: Last-Minute Entry Irks Other 2 Candidates," by Thomas B.

Pfankuch, which appeared in *(Jacksonville) Florida Times-Union* on August 27, 2002; "Citrus Interests Wage War on Barley," by Craig Pittman, which appeared in the *St. Petersburg Times* (now the *Tampa Bay Times*) on August 30, 2002; "Florida Farm Heavyweights Attack Ag Candidate," by Mary Ellen Klas, which appeared in the *Palm Beach Post* on August 30, 2002; "Bronson's Grip on Agriculture Post Faces Challenge," by Phil Long, which appeared in the *Miami Herald* on September 1, 2002; "Nelson Pulls Off Surprising Win—Attack Ads Undercut Mary Barley's Chances of Victory in the Primary for Agriculture Commissioner," by April Hunt, which appeared in the *Orlando Sentinel* on September 12, 2002; "Political Novice Counts on Canker Backlash," by Neil Santaniello, which appeared in the *Fort Lauderdale Sun Sentinel* on October 31, 2002; "Commissioner of Agriculture," by Craig Pittman, which appeared in the *St. Petersburg Times* on October 31, 2002; and "Republicans Sweep Cabinet with Wins by Crist, Bronson," by Beth Kassab, which appeared in the *Orlando Sentinel* on November 6, 2002.

CHAPTER 16. *A Big Deal*

The opening scene of Governor Charlie Crist's meeting with U.S. Sugar Corp. lobbyists came from "Big Sugar Took Its Lumps, Then Dealt," by John Kennedy and Aaron Deslatte, which appeared in the *Orlando Sentinel* on June 29, 2008. The scene of Crist's unveiling of the deal came from multiple sources, including interviews with Mary Barley and also "Saving the Everglades," an article I wrote with Arian Campo-Flores that appeared on Newsweek.com on August 19, 2008. Also helpful was "Sugar Buyout Hailed as Glades 'Gift'—A Landmark Deal Would Secure the 'Holy Grail' of Land to Help Revive and Clean Up the Everglades, but the Trade-Off Could Be Years of Additional Delays in Restoration Projects," by Curtis Morgan and Scott Hiaasen, which appeared in the *Miami Herald* on June 25, 2008. Background information on the deal came from "Few Knew Details of Sugar Buyout Talks—The $1.75 Billion State Plan to buy U.S. Sugar Was Hatched in a Tallahassee Lobbyist's Office Last Year," by Curtis Morgan and Scott Hiaasen, which appeared in the *Miami Herald* on June 26, 2008; "Sugar Deal Won't Let Water Flow—State Planners Say the U.S. Sugar

Property Will More Likely Be Used to Store Water Rather than Recreate a Flow from Lake Okeechobee to the Everglades," by Curtis Morgan and Scott Hiaasen, which appeared in the *Miami Herald* on June 29, 2008; and "2nd Firm Holds Key to Glades' Sweet Deal—If the State's Buyout of U.S. Sugar Succeeds, the Powerful Fanjul Sugar Barons Could End Up Holding the Key to Everglades Restoration," by Curtis Morgan and Scott Hiaasen, which appeared in the *Miami Herald* on July 4, 2008. Other significant sources were "Political Pluck, Power Dovetailed in State–U.S. Sugar Deal," by Stacey Singer, which appeared in the *Palm Beach Post* on June 29, 2008; and "Environmental Elites Are a Force behind Glades—The Little-Known but Well-Connected Florida Everglades Foundation Is a Major Player in the Restoration of the Florida Everglades," by Curtis Morgan, which appeared in the *Miami Herald* on July 27, 2008. To learn more about how residents of the Everglades Agricultural Area reacted to the news, I talked with Rick Roth and also referred to "Sugar Deal Puts Town's Fate in Limbo—As Clewiston Prepares for the Eventual Departure of U.S. Sugar, Local Views of the Future Range from Uncertain to Bleak, with a Few Optimistic Holdouts," by Jane Bussey, which appeared in the *Miami Herald* on July 10, 2008; and "In Florida, One Sugar Town's Bittersweet Change," by Jacqui Goddard, which appeared in the *Christian Science Monitor* on July 30, 2008. The anecdote about the reservoir came from "Reservoir Larger than Manhattan Planned to Help Everglades," by Brian Skoloff of the Associated Press, which appeared on Foxnews.com on May 6, 2008. A significant source on the deal's unwinding was "Deal to Save Everglades May Help Sugar Firm," by Don Van Natta Jr. and Damien Cave, which appeared in the *New York Times* on March 7, 2010. I also referred to "Land Deal Could Lift U.S. Sugar's Sagging Fortunes—Is It a Buyout or a Bailout? Either Way, a Pending Deal to Sell Land to the State for Everglades Restoration Could Reverse Big Sugar's Flagging Finances," by Jane Bussey and Curtis Morgan, which appeared in the *Miami Herald* on November 16, 2008; "Water District Taking $25M Hit for Halted Everglades Reservoir Work—Water Managers Have a Tentative Deal with a Contractor after Scuttling Work on a Reservoir," by Curtis Morgan, which appeared in the *Miami Herald* on September 10, 2009; "Charlie Crist's Incredible Shrinking Sugar Deal," by

Curtis Morgan, which appeared in the *Miami Herald* on March 7, 2010; and "Sugar Land Deal Finally a Lock—The Governor's Big Sugar Land Deal Is All but Completed, Though It's Not Nearly as Big as It Was When First Proposed More than Two Years Ago," by Curtis Morgan, which appeared in the *Miami Herald* on October 9, 2010. Also helpful was a fact sheet on the acquisition from the South Florida Water Management District.

CHAPTER 17. *Today's Big Picture*

A significant source on Burmese pythons in the Everglades was "Inside the Effort to Hunt Pythons Slithering Amok in the Florida Everglades," by Justin Worland, which appeared on Time.com on November 6, 2019. Also helpful was "South Florida Water Management District—Thousands from around the World Want to Be Python Hunters—The Burmese Python Is an Invasive Species in the Everglades; Florida's Python Elimination Program Will Add More Hunters, and Applications for the Position of Contracted Removal Agent Have Come from Far and Wide," by Jack Brook, which appeared in the *Miami Herald* on October 22, 2019. For alligators a primary source was "Scrawny Alligators Reflect Everglades' Many Problems," a story that I reported for WMFE (NPR Orlando) and that aired on January 21, 2016. Also helpful were "Everglades Alligators Wasting Away While Congress Controls Their Fate," by Craig Pittman, which appeared in the *Tampa Bay Times* on November 22, 2014. I also referred to articles on pythons, alligators, and crocodiles on *National Geographic*'s website, nationalgeographic.com, and the Florida Fish and Wildlife Conservation Commission website. I read and enjoyed "The Frail Future of an Alligator Hole," by Sasha Nyary and George Howe Colt, which appeared in *Life* magazine in September 1995. To write about the Picayune Strand and Comprehensive Everglades Restoration Plan I referred to the US Army Corps of Engineers website, which offers helpful information sheets on most aspects of CERP. I also referred to *The Everglades Handbook: Understanding the Ecosystem*, by Thomas E. Lodge. To write about the first years after CERP's implementation I relied on *The Everglades: Water, Bird, and Man*, which appeared in the *Economist* on October 6, 2005. Also helpful was "Projects Most Crucial to Success Have Been Delayed the Longest, a

Federal Report Says," by Craig Pittman, which appeared in the *St. Petersburg Times* (now the *Tampa Bay Times*) on July 3, 2007. Later information came from "Is Florida Moving Too Slow to Save Everglades? More than 15 Years into Restoration Work—The Halfway Mark for the Original Plan—the Everglades Remain Far from Fixed with the National Academies of Sciences Predicting That at the Current Pace the Job Will Take Another 100 Years to Complete," by Jenny Staletovich, which appeared in the *Miami Herald* on February 5, 2017. I also drew information from the National Academies of Sciences report itself. John Campbell of the US Army Corps of Engineers provided important data and context. I talked with Mary Barley about her more recent involvement in Everglades restoration. To write about periphyton I relied on an online article of the National Park Service and also the report "System-Wide Ecological Indicators for Everglades Restoration, 2018," which can be found on EvergladesRestoration.gov, a website offering a trove of information. To write about the Kissimmee River I referred to the US Army Corps of Engineers website and also "The Kissimmee: A River Re-curved," a story I reported for National Public Radio that aired on *Weekend Edition Sunday* on October 19, 2014 (https://www.npr.org/2014/10/19/356647396/the-kissimmee-a-river-recurved).

CHAPTER 18. *The Reservoir*

The opening summary came from news articles from the time, including my own reporting for WMFE (NPR Orlando) and also "Top Lawmaker Calls for Buying Up Sugar Land to Clean Everglades," by Mary Ellen Klas, which appeared in the *Miami Herald* on August 9, 2016; and "Buying Land to Save Water? It's Complicated," by Mary Ellen Klas, which appeared in the *Miami Herald* on January 6, 2017. The background about the reservoir that was repurposed into a shallower basin came from interviews with the US Army Corps of Engineers and South Florida Water Management District and also an information sheet on the district website. I relied on news articles to trace the chronology leading up to Negron's reservoir proposal, including "Will Florida Bay Survive the Summer?" by Jenny Staletovich, which appeared in the *Miami Herald* on June 19, 2016; "Florida Shores Yellow, Brown and Black All Over," by Jenny Staletovich, which appeared in the *Miami Herald* on February 28, 2016; and "Full-Page Everglades Ad Aims to Shame Florida Sugar Growers—A New Ad Calls U.S. Sugar 'Unwavering Obstructionist'; The Ad Also Suggests the Mott Foundation Committed to Conservation While Polluting Florida's Waters—U.S. Sugar Has Called the Letter 'Divisive Chatter,'" by Jenny Staletovich, which appeared in the *Miami Herald* on March 2, 2016. The anecdote about Central Marine came from my own reporting for the story "Whew! That Toxic Algae Smells Bad, Really Bad," which aired on WMFE (NPR Orlando) on July 21, 2016. To write about the 2017 legislative session I relied on "Florida Lawmakers Want to Buy $1.2 Billion of Farmland to Make Toxic Algae Plan Work," by Mary Ellen Klas, which appeared in the *Miami Herald* on January 26, 2017; "Is Florida Moving Too Slow to Save Everglades?—More than 15 Years into Restoration Work—the Halfway Mark for the Original Plan—the Everglades Remain Far from Fixed with the National Academies Of Sciences Predicting That at the Current Pace the Job Will Take Another 100 Years to Complete," by Jenny Staletovich, which appeared in the *Miami Herald* on February 5, 2017; "Sugar Growers to State: No Sale on Our Farmland South of Lake Okeechobee," by Mary Ellen Klas, which appeared in the *Miami Herald* on February 6, 2017; "Negron Offers Compromise on Everglades Reservoir—State-Owned and Leased Land," by Mary Ellen Klas, which appeared in the *Miami Herald* on April 4, 2017; and "Compromise Yields Gift for the Everglades: 78 Billion Gallons of Cleaner Water," by Mary Ellen Klas, which appeared in the *Miami Herald* on May 2, 2017. I also relied on interviews with Mary Barley and Rick Roth and my own reporting at the time for WMFE. The final section, about the termination of the option, came from my own reporting for WMFE and "South Florida Water Managers Cancel Lease Option, Ending Historic Deal to Buy Sugar Land," by Jenny Staletovich, which appeared in the *Miami Herald* on December 13, 2018.

CHAPTER 19. *What Is Restoration?*

To envision a restored Everglades of the future I talked with Mary Barley and Rick Roth. Another significant source was "Synthesis of Everglades

Research and Ecosystem Services," SERES Project, accessed April 26, 2020, http://www.everglades-seres.org/SERES-_Everglades_Foundation/Welcome.html. The summary on peat soil came from the SERES Project report and also an article on Everglades geology on the National Park Service website (https://www.nps.gov/ever/learn/nature/evergeology.htm). To write about climate change and sea level rise I relied on a 2018 report from the National Academies of Sciences. Other significant sources were "Sea Rise Is Outpacing Everglades Restoration—But Scientists Say There's a Solution," by Jenny Staletovich, which appeared in the *Miami Herald* on February 12, 2018; and "Climate Change Throws a Wrench in Everglades Restoration," by Kate Stein, which appeared in *Scientific American* on June 21, 2019. George Barley's own writings were part of Mary Barley's personal records.

CHAPTER 20. *Looking-Glass Water*

I wrote the scene of Florida Bay after a boat tour with Jim Fourqurean, Tom Frankovich, and Mary Barley. Also helpful was "Is Florida Moving Too Slow to Save Everglades?—More than 15 Years into Restoration Work—the Halfway Mark for the Original Plan—the Everglades Remain Far from Fixed with the National Academies of Sciences Predicting That at the Current Pace the Job Will Take Another 100 Years to Complete," by Jenny Staletovich, which appeared in the *Miami Herald* on February 5, 2017. My summary of the 2018 gubernatorial race came from my own reporting for WMFE (NPR Orlando) and other news articles, including "This Candidate for Florida Governor Is the Only One Taking Money from Big Sugar," by Craig Pittman, which appeared in the *Miami Herald* on August 6, 2018; "DeSantis Rolls Out Environmental Platform As Critics Call Him 'Sham Environmentalist,'" by Martin Vassolo and David Smiley, which appeared in the *Miami Herald* on September 12, 2018; and "Everglades Trust Endorses Ron DeSantis over Andrew Gillum," by Adam C. Smith, which appeared in the *Miami Herald* on October 16, 2018. Also helpful were "Democrats Talk Lake O Runoff Ills, School Risk—Gubernatorial Candidates Slam Scott, Trump in Debate," by Lisa Conley, which appeared in the *Naples Daily News* on July 19, 2018; and "Democrats' Debate—Algae an Issue in Race for Governor," by Ali Schmitz, which appeared in the *Naples Daily News* on August 4, 2018. I also talked with Mary Barley. I referred to state campaign contribution records for the amount Paul Tudor Jones gave to DeSantis's campaign. To reconstruct DeSantis's first year in office I again relied on my own reporting for WMFE and also news articles, including "DeSantis calls for Water Managers to Resign, Pulls Scott Appointments," by Jenny Staletovich, which appeared in the *Miami Herald* on January 11, 2019; "DeSantis Unveils Sweeping Fixes Meant to Address Water Woes: In His First Significant Policy Move, Gov. Ron DeSantis on Thursday Visited Areas Hardest Hit by Algae Blooms and Red Tide to Announce Measures to Fix Florida's Troubled Waters," by Jenny Staletovich, which appeared in the *Miami Herald* on January 11, 2019; and "Is Ron DeSantis Really Florida's Green Governor? We're About to Find Out," by David Smiley and Adriana Brasilerio, which appeared in the *Miami Herald* on May 8, 2019. The anecdote about the community meeting in Islamorada came from "Water Plan Leaves Bay Advocates Parched," by Kevin Wadlow, which appeared on Keysnews.com on February 26, 2020. To write about the C-111 Spreader Canal Western Project, I relied on "Signs of New Life: Water Flow, Plant Life Increase in First Year of Project's Operation," by Sue Cocking, which appeared in the *Miami Herald* on January 22, 2014; and "Dry winter, Slow Progress on Everglades Work Puts Florida Bay at Risk," by Jenny Staletovich, which appeared in the *Miami Herald* on May 3, 2015. Also helpful was a US Army Corps of Engineers information sheet.

INDEX